THE ECONOMICS OF CONTRACT CHOICE

The Economics of Contract Choice

An Agrarian Perspective

YUJIRO HAYAMI
AND
KEIJIRO OTSUKA

CLARENDON PRESS·OXFORD
1993

Oxford University Press, Walton Street, Oxford OX2 6DP
Oxford New York Toronto
Delhi Bombay Calcutta Madras Karachi
Petaling Jaya Singapore Hong Kong Tokyo
Nairobi Dar es Salaam Cape Town
Melbourne Auckland
and associated companies in
Berlin Ibadan

Oxford is a trade mark of Oxford University Press

Published in the United States
by Oxford University Press, New York

© Yujiro Hayami and Keijiro Otsuka 1993

All rights reserved. No part of this publication may be reproduced,
stored in a retrieval system, or transmitted, in any form or by any means,
electronic, mechanical, photocopying, recording, or otherwise, without
the prior permission of Oxford University Press

British Library Cataloguing in Publication Data
Data available

Library of Congress Cataloguing in Publication Data
Hayami, Yūjirō, 1932–
The economics of contract choice : an agrarian perspective /
Yujiro Hayami and Keijiro Otsuka.
p.   cm.
Includes bibliographical references (p.   ) and index.
1. Contracts, Agricultural—Developing countries. I. Otsuka,
Keijiro. II. Title.
HD1417.H29 1991   333.33'55'091724—dc20   91-27291
ISBN 0-19-828378-4

Typeset by Alliance Phototypesetters, 50, 7th Cross, Brindavan, Pondicherry
Printed in Great Britain by
Bookcraft (Bath) Ltd,
Midsomer Norton, Avon

To D. Gale Johnson

Preface

THIS book is a synthesis of our research over nearly two decades into contract choice in agrarian economies. The research began in 1974 when the senior author was intrigued by numerous informal contracts or agreements governing production and resource use in a small rice village in the Laguna Province of the Philippines, for which he undertook a survey for a project of the International Rice Research Institute (IRRI). For one who was then unaccustomed to villages in the Third World, these contracts among villagers, apparently quite complicated and variable but seldom written or even orally confirmed, were elusive and difficult to understand. He continued his research on foot into the pattern and the mechanism of contract choice and enforcement in peasant communities as his field-work was expanded to major areas of the Philippines as well as other countries. (The results were published in Y. Hayami and M. Kikuchi, *Asian Village Economy at the Crossroads*, University of Tokyo Press and Johns Hopkins University Press, 1982.)

Later, his interest was transmitted to the younger author. He had been interested in curious features of share tenancy since his acquaintance with this subject in the graduate school of economics at the University of Chicago in the mid-1970s. He examined the problem of the land and labour contracts in agrarian economies based on the modern economic theory of contract or the theory of principal and agent relationship.

Our perspective has evolved through dialogue between theory and empirical observation. Fortunately we were able to share many interesting field observations through joint participation in a recent IRRI project, entitled 'Impact of the Modern Rice Technology across Different Production Environments', covering seven countries in South and South-East Asia. Preparation of the manuscript was facilitated by support from the Center for International Studies in the Research Institute of Aoyama-Gakuin University.

This book draws extensively on our earlier publications, including those with other co-authors, as acknowledged in text and notes. Especially important are the contributions from three collaborators: Hiroyuki Chuma for theoretical developments in Chapters 2–5; and Masao Kikuchi and Yoshinori Morooka for case-studies in Chapters 7 and 8 respectively. Two of our earlier papers (K. Otsuka and Y. Hayami's 'Theories of Share Tenancy: A Critical Survey', *Economic Development and Cultural Change*, 1988; K. Otsuka, H. Chuma, and Y. Hayami's 'Land and Labor Contracts in Agrarian Economies: Theories and Facts', *Journal of Economic Literature*, 1992) provided the framework for this volume.

We have benefited from comments and suggestions at various stages of research from Franklin Allen, Lee Alston, Randolph Barker, Hans Binswanger, Avishay Braverman, Martin Broffenbrener, Michael Carter, Cristina David, Alain de Janvry, Jose Encarnacion, Robert Evenson, Akimi Fujimoto, Koichi Hamada, Mahabub Hossain, Shigeru Ishikawa, Motoshige Itoh, Yoshitsugu Kanemoto, Toshihiko Kawagoe, Ashok Kotwal, Justin Lin, Naoki Murakami, Keiichiro Nakagawa, Koji Nishikimi, Konosuke Odaka, Hiroshi Ohta, Masahiro Okuno, John Pencavel, Gustav Ranis, C. H. Hanumantha Rao, Sherwin Rosen, Mark Rosenzweig, Vernon Ruttan, Theodore Schultz, T. N. Srinivasan, and Anthony Tang. We are also grateful, for assistance in collecting and processing data, to Yolanda Aranguren, Luisa Bambo, Violeta Cordova, Esther Marciano, Henny Mayrowani, Milagros Obusan, and Dolor Palis. For all these benefits we express our deepest gratitude.

This book is dedicated to D. Gale Johnson. His personal stimulus together with his path-breaking study of agricultural tenancy ('Resource Allocation under Share Contracts' *Journal of Political Economy*, April 1950), though not as well known as it should be, has been an invaluable guide-post for our work.

YUJIRO HAYAMI
KEIJIRO OTSUKA

Contents

List of Figures   xiii

List of Tables   xiv

1. Agrarian Organizations and Contracts   1

    1.1 Issues and Focus   2
    1.2 The Setting of Agrarian Economies   6
    1.3 Plan of the Book   18

2. The Basic Model   21

    2.1 Specification of Functional Forms   22
    2.2 The Formulation of the Optimization Problem   26

3. Optimum Contract Choice under Alternative Assumptions   34

    3.1 Optimum Contract Choice under 'Certainty'   34
    3.2 Can Share Contract Exist under 'Certainty'?   39
    3.3 Optimum Contract Choice under Uncertainty   46
    3.4 On the Relevance of Alternative Assumptions   51
    Appendix: Derivation of the Optimum Conditions   52

4. Long-Term Contracts   56

    4.1 Theories of Long-Term Contracts   57
    4.2 Theories of Long-Term Agrarian Contracts   59
    4.3 Reputation and Contract Enforcement   64

5. Interlinked Contracts   70

    5.1 Interlinked Credit Contracts   71
    5.2 Cost-Sharing Contracts   78
    5.3 On the Advantage of Interlinked Contracts   82

Contents

6. Global Survey of Empirical Evidence — 85
   - 6.1 Conditions of Permanent-Labour Contracts — 85
   - 6.2 Efficiency and Income Distribution under Share Tenancy — 89
   - 6.3 Mechanism of Effective Share-Contract Enforcement — 98
   - 6.4 Consequences of Tenancy Regulations — 101
   - 6.5 Spectrum of Contract Choice — 103

7. Contract Choice and Enforcement in an Agrarian Community: The Case of Upland Farming in Indonesia — 107
   - 7.1 Characteristics of the Study Village — 109
   - 7.2 Land-Tenure Systems — 113
   - 7.3 Relative Efficiencies and Enforcement Costs — 117
   - 7.4 Community Mechanism of Contract Enforcement — 123
   - 7.5 Conclusion — 126

8. Community and Market in Contract Choice: The Case of the Jeepney in the Philippines — 128
   - 8.1 Jeepney Operations and Hypotheses — 129
   - 8.2 Testing the Alternative Hypotheses — 135
   - 8.3 Conditions of Effective Contract Enforcement — 141
   - 8.4 Conclusion — 144
   - Appendix: Estimation of Elasticity of Substitution — 145

9. Land Reform, New Technology, and Agrarian Contracts: The Case of a Philippine Rice Bowl — 148
   - 9.1 Study Site and Data Collection — 148
   - 9.2 Changes in Labour Relations — 150
   - 9.3 Conditions of New Permanent-Labour Contracts — 154
   - 9.4 Tests on Relative Income and Efficiency of Permanent Labour — 159
   - 9.5 Conclusion — 170

10.	Towards a General Theory of Agrarian Contracts	172
	10.1 Theoretical Conclusions	172
	10.2 Empirical Conclusions	175
	10.3 Tasks Ahead	176

References 179

Index 201

List of Figures

3.1	Model of contract choice under 'certainty'	37
3.2	Self-selection model of contract choice	42
3.3	The land-quality transaction-cost model of contract choice	46
6.1	Frequency distribution of the rates of difference in output per hectare of share-tenancy land from that of owner-farming and leasehold-tenancy land	93
7.1	The Garut District in West Java, Indonesia	111
7.2	Average monthly rainfall and a typical cropping system in the study area	112
8.1	Distribution of jeepneys by form of rental contract in the Southern Tagalog Region, Philippines, 1982	132
9.1	Geographical distribution of different types of permanent-labour arrangements in twenty-eight municipalities, Central Luzon, August 1987 (based on the Permanent-Labour Contract Survey)	152

List of Tables

1.1	Distribution of farms and farmland by land-tenure status in major regions of the world according to the 1970 World Census of Agriculture	8
1.2	Distribution of farms and farmland by operational farm size and land-tenure status in selected developing countries in Asia and Latin America	10
3.1	Relative magnitudes of output, input, land rent, and tenant income from farming among three land-tenure classes, implied from different assumptions about contract enforcement	39
6.1	Frequency distribution of the rates of difference in output per hectare of share tenancy land from that of owner-farming and fixed-rent tenancy land in past case-studies	91
6.2	Average rates of difference in labour and fertilizer inputs per hectare of share-tenancy land from those of owner-farming and fixed-rent tenancy land in past case-studies	95
7.1	Size distribution of land ownership and operational holdings by villagers in the study village, Garut, Indonesia, 1986	110
7.2	Distribution of farm land, farm-operators, and landlords by tenure status in the study village, Garut, Indonesia, 1986	113
7.3	Distribution of cost- and output-sharing arrangements for soybean, corn, and tobacco by plot under share tenancy (*maro* and *mertilu*) in the study village, Garut, Indonesia, 1986	115
7.4	Characteristics of landlords and tenants by tenancy contract in the study village, Garut, Indonesia, 1986	116
7.5	Average values of gross output, gross value added, current input, and labour input per hectare by tenure class in the study village, Garut, Indonesia, 1986	118
7.6	Distribution of plots among different cropping systems for different land-tenure classes in the study village, Garut, Indonesia, 1986	119
7.7	Comparison of estimated share rent with fixed rent	122
7.8	Frequency of landlords' visits to rented-out plots	124
7.9	Personal relations between landlords and tenants	125

List of Tables

7.10	Cumulative duration of tenancy contracts	126
8.1	Orderings in the magnitudes of output, input, rental, and driver income from jeepney operations between the share and fixed-rental contracts, predicted from the three alternative hypotheses	135
8.2	Average revenue, jeepney rental, and driver income per jeepney per day in a sample of jeepney routes classified by dominant form of contract in the Southern Tagalog region, Philippines	136
8.3	Average revenue, value added, labour, and current inputs per jeepney per day by form of contract on the Calamba–San Pablo route, Laguna, Philippines	138
8.4	Estimation of average depreciation and repair costs of jeepneys by form of contract on the Calamba–San Pablo route, Laguna, Philippines	139
8.5	Distribution of value added per day from jeepney-operation among owners and drivers by form of contract on the Calamba–San Pablo route, Laguna, Philippines	140
8.6	Socio-economic characteristics of jeepney-owners and drivers by form of contract on the Calamba–San Pablo route, Laguna, Philippines	142
8.7	Estimates of the elasticity of substitution function	146
9.1	Changes in rice technology, land-tenure distribution, and the incidence of permanent labour in Central Luzon and the Nueva Ecija village	153
9.2	Matrix of land transfer through pawning of CLT and leasehold titles in the Nueva Ecija village, 1987	158
9.3	Socio-economic characteristics of sample farmers, permanent labourers, and casual labourers in the Nueva Ecija village, 1989 dry season	161
9.4	Socio-economic characteristics of employers of permanent labourers in the Nueva Ecija village, 1989 dry season	162
9.5	Labour income and workdays of permanent and casual labourers in the Nueva Ecija village by source, 1989 dry season	163
9.6	Comparison of labour inputs per hectare between the farms of farmers' family-labour operation and of permanent-labour operation in the Nueva Ecija village, 1989 dry season	164
9.7	Comparison of average output value, production costs, and residual profit per hectare between farms with and without permanent labour in the Nueva Ecija village, 1989 dry season	167

1

Agrarian Organizations and Contracts

THE basic tenet of neoclassical economics is that competitive markets can achieve efficient resource allocation (in the Pareto-optimum sense) where information is perfect and a complete set of markets exists. To the extent that information is imperfect, transaction costs associated with market exchanges may become so high that the market needs to be replaced by other forms of institution and organization. In order to meet this demand, especially to save the cost of monitoring labourers' work-effort, hierarchical internal organizations called 'firms' have emerged in modern urban economies (Coase 1937; Alchian and Demsetz 1972; Williamson 1975, 1985). In agrarian economies, however, small-scale family farms have continued to be the dominant mode of production, not only in low-income developing countries but also in high-income industrial countries, even though the operational farm size in the latter is usually much larger than in the former.

It is commonly observed that population in agrarian economies is stratified into peasant subclasses ranging from landless labourers to non-cultivating landlords. Various arrangements of land-tenancy and labour-employment contracts are used for combining land and labour for agricultural production in the decentralized system of small-scale family farms. Why is it that hierarchical firms do not become dominant in agrarian economies? What contractual arrangements among independent farm-operators, labourers, and landlords are substituted for the large internal organizations? Why is one form of contract chosen among alternatives in the economic, social, and technological environments of agrarian economies? What is the mechanism of enforcing contractual terms, especially with respect to the worker's effort? These are the major questions addressed in this book.

While our analysis is mainly intended to identify the characteristics of land and labour contracts in agrarian economies, the results are also expected to produce useful insights into the nature of the organization of production in urban economies in contrast with that of agrarian economies.

1.1 Issues and Focus

Land-tenancy and labour-employment contracts are alternative arrangements to combine the two primary factors of production in agrarian economies. The choice between these alternatives has been discussed in the economic literature on agrarian organization from the classic theory of the agricultural ladder by Spillman (1919) to a recent synthesis by Binswanger and Rosenzweig (1986). Yet, in formal economic models, land and labour contracts have been treated separately. The prolific literature on land tenancy has focused on the choice between share and fixed-rent tenancy contracts (Newbery and Stiglitz 1979; Binswanger and Rosenzweig 1984; Quibria and Rashid 1984; Otsuka and Hayami 1988; Singh 1989), while studies on agrarian labour relations have focused mainly on the choice between casual- and permanent-labour contracts (Bardhan 1979*b*, 1983; Eswaran and Kotwal 1985*a*; Bell and Srinivasan 1985*b*) or the choice among different remuneration systems for casual labour such as time-rates and piece-rates (Roumasset and Uy 1980). However, separate analyses of land and labour contracts have resulted in much theoretical confusion as well as questionable interpretations of empirical data, especially those drawing inferences about the relative efficiency of alternative contracts. This study aims to point toward a 'general theory of agrarian contracts', in which land tenancy and labour employment are modelled together as substitutes along a continuous spectrum of contract choice, through a critical review of existing studies of agrarian contracts and our own empirical analyses.

Partial models of land tenancy can explain some basic stylized facts; among others (*a*) the pervasiveness of share-cropping in history and in many different places and (*b*) the decline in its incidence in more developed countries (Stiglitz 1989). Moreover, partial models of labour employment could explain the two-tiered structure of the agrarian labour market in which permanent or attached labourers employed for a crop season or longer receive higher remuneration than casual labourers employed on a daily basis (Bardhan 1979*b*, 1983; Eswaran and Kotwal 1985*a*). However, as these models do not deal with the choice between land and labour contracts, they cannot answer the fundamental question of why family farms of either owner- or tenant-operation have continued to be a more frequent mode of agricultural production than large-scale farm firms or plantations based on hired labour, not only in low-income developing economies but also in high-income industrial economies. Nor can they explain major differences between regions in the distribution of land and labour contracts, e.g. the incidence of permanent-labour contracts is relatively more common in South Asia, especially India, than in South-East Asian countries such as Indonesia, the Philippines, and Thailand. A broader and more convincing understanding of agrarian organizations in the Third World is unlikely to be achieved unless a general theory of land and labour contracts is developed.

Our unified treatment of land and labour contracts is based on the theory of the principal–agent relationship or, briefly, agency theory (Arrow 1985; Hart and Holmstrom 1987; Levinthal 1988). Agency theory is concerned with designing an optimum contract between a principal (owner of a resource) and an agent (user of the resource). An agency problem arises when the agent's action (e.g. work-effort) is not directly observable by the principal and the outcome is influenced not only by the agent's action but also by uncertain factors outside the agent's control. In this situation the principal cannot verify whether a poor outcome is due to the agent's idleness or to an unfavourable state of nature. Thus the agent tends to work less hard than the principal would like. If, however, the agent's income is made to depend in some fashion on the consequences of his effort, the agent has an incentive to carry out his work efficiently. The defect with such an arrangement is that it renders the agent's income variable because the product of his work depends not only on his own action, but also on the factors beyond his control. Being risk-averse, the agent prefers a certain income to an uncertain one. The optimum contract seeks a balance between providing work-incentives for the agent and lessening his exposure to risk.

A forerunner of modern agency theory was the analysis of land tenancy. More than a century ago, classical economists like Adam Smith (1776) and John Stuart Mill (1848) analysed the relative merits of fixed-rent leasehold tenancy prevalent in England and share tenancy (*métayage*) prevalent in France. Their analysis emphasized the different investment- and work-incentives for tenants under alternative contracts (Cheung 1969; Bliss and Stern 1982; Jaynes 1984). Later, Alfred Marshall (1890) formalized the efficiency implications of share vs. fixed-rent contracts. The dominant view from Marshall maintained that share tenancy resulted in inefficient resource allocation because the share-tenant received as marginal revenue only a fraction of the marginal product of his labour and this lessened his incentive to work (Schickele 1941; Heady 1947; Castle 1952; Drake 1952; Issawi 1957; Georgescu-Roegen 1960; Bhagwati 1966; A. K. Sen 1966; Adams and Rask 1968).

However, Marshall himself and, later Johnson (1950) and Cheung (1969) argued that, if the tenant's work-effort could be monitored without cost and enforced by the landlord, resource allocation under share tenancy could be as efficient as under owner-farming and fixed-rent tenancy. The requirement of costless monitoring, however, has been criticized as unrealistic (Bardhan and Srinivasan 1971; Stiglitz 1974; Mazumder 1975; A. K. Sen 1975; Bell and Zusman 1976; Bardhan 1977, 1979a; Bell 1977). If the tenant's effort is costly to monitor, then the share contract may be rationalized as a risk-sharing device (Stiglitz 1974).

The major controversy has centred around the enforceability of contractual terms. It is difficult for a landlord to monitor a tenant at work at various farm tasks throughout the season in a dispersed and ecologically diverse agricultural environment. On the other hand, a small agrarian community in developing economies is typically characterized by close personal relations between landlord and tenant, and the landlord may be able to collect sufficient information regarding his tenant's conduct by casual observation and gossiping without undertaking explicit monitoring activities. These personal relationships encourage the development of long-term contracts interlinked with credit and insurance agreements that tend to increase the costs of terminating a tenant's contract in the event of possible discovery of opportunistic or dishonest behaviour. The controversy on contract enforceability may be settled by understanding rural community relations that reinforce the mechanism of contract enforcement envisaged by modern theories of incentive-contracting based on the threat against the agent of explicit penalty payments (e.g. Mirrlees 1974; Lewis 1980; Radner 1981; Rubinstein and Yaari 1983).

A major confusion in the literature on agrarian institutions stems from faulty assumptions regarding the enforceability of land and labour contracts. Tenant work-incentives are weaker under share than fixed-rent tenancy because the share-tenant can claim only a fraction of the residual profit. So it should follow that labourers hired at a fixed wage-rate have even weaker work-incentives and that the enforcement of their work-effort poses an even more serious problem than in the case of share-tenants. This problem of incentives is considered a key to understanding the dominance of small-scale family farms in agriculture. In the Third World, large-scale farm firms or plantations based or. hired labour under central management are mainly engaged in the production of cash crops for export where scale economies arise from the need to co-ordinate production with bulk shipment and/or industrial processing, while small traditional peasant-farming is predominant in the production of major food crops. Even in the production of cash crops for export, small farms where the peasants are effectively linked with agribusiness marketing and processing firms through production and delivery contracts have often proved more efficient than large plantations (W. A. Lewis 1969; Hayami et al. 1990). In socialist countries, the relative inefficiency of the agricultural sector has been ascribed to the difficulty of supervising labourers in collective farms (Bradly and Clark 1972; Lin 1988). Yet in theories of land and labour contracts, it is not uncommon to assume that share-tenancy contracts are difficult to enforce whereas wage-labour contracts are costlessly enforced, e.g. compare the labour-contract models of Bardhan (1979b; 1983) with the share-tenancy models of Bardhan and Srinivasan (1971) and Bardhan (1977, 1979a). Such contradictory assumptions have led to the conclusion that there is a 'dominance of wage labour over land tenancy contract' (Bell and Braverman 1980), a proposition that is totally incompatible with the actual pattern of agrarian organization in the Third World. Inconsistent assumptions with respect to the enforceability of work-effort have resulted from separate piece-meal treatment of land and labour contracts that have marred a real understanding of why small farms predominate in agrarian economies. We attempt to answer this fundamental question while sharing basically the same approach as that of the modern theory of industrial organization in terms of its emphasis on risk, transaction costs, and incentives (Coase 1937; Alchian and Demsetz 1972; Arrow 1974; Williamson 1975, 1985). Our analysis is relevant not only to agrarian economies but also to many sectors of urban economies (such as commerce, professional and personal services) in which small self-employed enterprises have a comparative advantage, but the major purpose of this book is to search for a model of land and labour contracts based on general and consistent assumptions.

1.2 The Setting of Agrarian Economies

The 'agrarian economies' we are concerned with in this book are those of the peasant sector in the Third World. In this sector a large number of small-scale family farms are operated for the family's subsistence by producing mainly staple food crops for home consumption as well as for sale, with cash income supplemented by sideline production of commercial crops and livestock. Even today this mode of peasant production provides a livelihood for the majority of mankind. We exclude from our analysis large-scale farms or plantations specializing in the export of tropical cash crops. Analysis of the organization of production by such agribusiness plantations would require little additional explanation beyond the established theory of internal organizations or internal labour markets in urban economies, although the issue of possible advantages of scale arising from power concentration corresponding to highly skewed land-ownership distribution, such as in Latin America, may require separate investigation. Moreover, our analysis is limited to situations in which private property rights in land are established formally or informally because in the economies where tribal or communal land-tenure prevails contract choice by individual producers does not apply. The chief characteristics of the agrarian economies we are concerned with are summarized below.

Agrarian structures

First, we compare agrarian structures in major regions of the world in terms of the distribution of farms and farmland by both land-tenure status and farm size. The data are from the 1970 World Census of Agriculture compiled by the United Nations Food and Agriculture Organization. Needless to say, the reliability of these data differs greatly among countries, while a number of countries are not covered, and, in the case of Africa, the countries covered are not very representative. Yet broad comparisons suggest some important characteristics of agrarian structures in the world.

As can be seen from Table 1.1, world-wide, owner-cultivation is the most common form of land tenure comprising 79 per cent of all farms and 61 per cent of total farmland in all the forty-six countries covered by the 1970 World Census. Note the especially high incidence of owner-cultivation in Asia together with the small size of the average farm in this region (2.3 ha.). This reflects the fact that Asian agriculture is dominated by 'peasants'—as defined by Chayanov (1966)—who own small landed properties and operate them on a family basis. This peasant mode is augmented by the use of tenancy contracts that facilitate land transfers from relatively land-abundant households to households with little land so as to make the ratio of farmland to family labour more or less equal across farms. Permanent-labour contracts that facilitate the development of large farm firms through labour transfers from land-scarce (typically landless) households to large landowners are less common as a means to equalize the ratio of land to labour.

Owner-cultivation persists in advanced industrial economies such as Europe and North America. Average farm sizes in these regions are much larger than in Asia. In part, this reflects differences in environmental conditions and cropping systems. More importantly, it indicates the much higher opportunity costs of farm labour because larger operational holdings are needed to earn sufficiently high incomes for family members to stay on the farm (Kislev and Peterson 1982). The greater incidence of tenancy in Europe and North America than in Asia reflects the pressure for farm size adjustments corresponding to rapid absorption of farm labour by non-agricultural sectors, together with scale economies arising from substitution of increasingly large-scale farm machinery for labour. As shown in Table 1.1, the incidence of share tenancy is significant in these regions, especially in North America, though lower than in Asia, suggesting that it plays an important role in facilitating farm size adjustments not only in developing but also in developed economies.

A major exception to the general dominance of owner-cultivation is Africa where as many as 86 per cent of farms representing 59 per cent of farmland are operated under forms of tribal or communal tenure (included in the 'other' tenure category in Table 1.1). In this respect, Latin America is closer to Asia even though squatters

TABLE 1.1. *Distribution of farms and farmland by land-tenure status in major regions of the world according to the 1970 World Census of Agriculture*

	Asia	Africa	Latin America	Europe	North America	World
Number of countries enumerated[a]	10 (5)	4 (1)	15 (7)	12 (3)	2 (1)	46 (17)
Number of farms (million)	93.3	3.5	8.6	11.9	3.1	120.4
Average operational farm size (ha.)	2.3	0.5	46.5	7.6	161.2	10.0
Distribution of farms (%):						
Owner-cultivation	85.8	5.2	60.3	67.6	63.2	79.2
Pure tenancy	5.9	1.6	17.1	9.3	12.0	7.1
Owner-cum-tenancy	8.2	6.9	6.6	23.0	24.8	10.0
other[b]	0.	86.3	16.0	0.1	0.	3.7
Distribution of farmland (%):						
Owner-cultivation	84.0	9.2	80.4	58.9	36.6	61.1
Pure tenancy	5.9	3.0	6.2	12.5	11.9	9.0
Owner-cum-tenancy	10.1	29.1	5.6	28.5	51.5	27.2
Other[b]	0	58.7	7.8	0.1	0	2.7
Percentage of share tenancy in tenanted land	84.5	0.	16.1	12.5	31.5	36.1

[a] Countries enumerated are: *Asia*: Bahrain*, India, Indonesia, Jordan*, Korea, Kuwait*, Pakistan*, Philippines*, Saudi Arabia, Singapore; *Africa*: Cameroon, Réunion, Swaziland; *Latin America*: Costa Rica*, Dominican Republic*, El Salvador*, Guadeloupe*, Honduras, Panama, Puerto Rico, St Lucia, Virgin Islands, Brazil, Columbia*, Peru*, Surinam, Uruguay*, Venezuela; *Europe*: Austria, Belgium, France*, West Germany, Italy, Malta*, Netherlands, Norway, Poland, Portugal*, Sweden, UK; *North America*: Canada, USA*; *World* includes American Samoa, Guam, and Pacific Islands in addition to the above cited 43 countries. Countries with asterisks are those for which the data on share tenancy are available (number indicated in parentheses).

[b] Farms operated by squatters and under tribal on communal tenure forms.

Soucre: UN-FAO, *1970 World Census of Agriculture, Analysis and International Comparison of the Results.* Rome, 1981.

(also included in the other category) represent a significant share of the land-tenure distribution. However, the size distribution of farms in Latin America is totally different from that of Asia (Table 1.2). In Asia more than 70 per cent of farms are smaller than 5 ha. and they represent 40 to 70 per cent of farmland; land organized in large farms above 50 ha. is typically less than 10 per cent of total farmland. In contrast, in Latin America, small farms below 5 ha. constitute less than 10 per cent of total farmland, while more than 70 per cent of land is held by large farms above 50 ha. Correspondingly, the Gini coefficient measuring inequality in the distribution of farmland is higher than 0.8 for Latin American countries as compared with less than 0.6 for Asian countries. These data reflect the fact that most agricultural production in Asia involves large numbers of peasants organized around family farms, whereas in Latin America most agricultural production takes place on large plantations or haciendas using hired labour. This contrast derives from the different historical experiences of these two areas (Hayami *et al.* 1990, ch. 1).

These structural differences have given rise to different agrarian problems among the three major regions of the Third World. In Africa the chief concern has been how to develop modern property rights in land consistent with optimum production, investment, and resource-conservation incentives (Clayton 1964; Lower 1986); and in Latin America it has related to the transformation of semi-feudal estates based on bonded labour into capitalist farm firms based on hired labour (de Janvry 1981). The traditional agrarian problem in Asia has been concerned with the landlord–tenant relationship (Ladejinsky 1977). Of course, these are sweeping generalizations. In some parts of Africa and Latin America the landlord–tenant relationship is also a serious problem (de Janvry 1981; Robertson 1987). In Asia, too, especially in the Philippines, the conflict between owners of plantations and labourers is a critical issue (Hayami *et al.* 1990). The problem of how to develop and assign modern property rights in land is also an important issue on the policy agenda of the Third World in general (Bromley 1986). Yet by far the majority of existing studies on land tenancy relate to the Asian context.

Within Asia, farm size distributions differ among countries owing to both environmental and historical conditions. Large farms above 50 ha. are relatively frequent in Indonesia and the Philippines

TABLE 1.2. *Distribution of farms and farmland by operational farm size and land-tenure status in selected developing countries in Asia and Latin America*

Country	Year of survey	Average operational farm size (ha.)	Percentage of farms and farmland				Gini coefficient of land concentration	Percentage of tenanted area in total farmland		Percentage of share tenancy in tenanted land[b]
			Below 5 ha.		Above 50 ha.			pure tenancy	Total[a]	
			Farms	Area	Farms	Area				
Asia										
Bangladesh	1976/7	1.6	90.6	67.6	n.a.	n.a.	.42	n.a.	20.9	91.0
India	1970/1	2.3	88.7	46.7	0.1	3.7	.62	2.4	8.5	48.0
Indonesia	1973	1.1	97.9	68.7	0[c]	13.6	.56	2.1	23.6	60.0
Nepal	1971/2	1.0	97.2	72.1	0[c]	0.8	.56	1.5	13.2	48.3
Philippines	1971	3.6	84.8	47.8	0.2	13.9	.51	21.4	32.8	79.3
Thailand	1978	3.7	72.3	39.4	0	0.9	.45	6.0	15.5	29.0

TABLE 1.2 (cont.)

Latin America								
Brazil	1970	59.7	36.8	1.3	16.3	84.6	.84	6.1	10.2	n.a.
Costa Rica	1973	38.1	48.9	1.9	14.5	79.7	.82	1.2	9.0	9.4
Columbia	1970/1	26.3	59.6	3.7	8.4	77.7	.86	5.3	11.5	49.4
Peru	1971/2	16.9	78.0	8.9	1.9	79.1	.91	4.5	13.6	0[c]
Uruguay	1970	214.1	14.3	0.2	37.6	95.8	.82	19.1	46.3	4.7
Venezuela	1971	91.9	43.8	0.9	13.6	92.5	.91	4.5	2.4	n.a.

[a] Area in 'pure' tenant farms plus area in owner-cum-tenant farms.
[b] Percentage in the area of 'pure' tenant farms, except percentage in total tenanted area for Bangladesh.
[c] Less than 0.05%.
n.a. not available.

Sources: UN-FAO, *1970 World Census of Agriculture, Analysis and Comparison of the Results*, Rome, 1971; Government of Bangladesh, *Report on the Agricultural Census of Bangladesh, 1977*, Dhaka, 1981; *1978 Land Occupancy Survey of Bangladesh*, Dhaka; Government of India, *National Sample Survey, No. 215, 26th Round; 1971–72*, Tables on Land Holdings, New Delhi, 1976; Government of Indonesia, *1983 Agricultural Census*, Jakarta, 1986; Government of Nepal, *1981/82 National Sample Census of Agriculture*, Kathmandu, 1987; Government of Thailand, *1978 Agricultural Census Report*, Bangkok (publication year not indicated).

which have comparative advantage in plantation crops promoted by colonial governments. Moreover, the incidence of tenancy varies across countries, owing much to different political histories: the high incidence of tenancy in the Philippines is based on the Spanish colonial policy akin to that applied to Latin America; the low incidence of tenancy in India seems to reflect the effects of land reform after independence (both actual land redistribution and census underreporting); and the high incidence of share tenancy in Bangladesh has resulted from the prohibition of fixed-rent tenancy. These cases will be discussed in more detail in Chapter 6.

Analysis in this book focuses on the causes and consequences of contract choice by peasant producers in the Third World based on their private decisions. Although the peasantry is a very significant segment of the rural sector in Africa and Latin America, it is much more dominant in Asia than in other regions as well as in other sectors in this region. Therefore, it is natural that Asia should be the focus of this study. The next two subsections explain the technological, economic, and social environments under which these peasants are operating. We focus on how these environments condition the risks and transaction costs that underlie peasants' contract choices.

Production environment

Agricultural production is characterized both by uncertainty regarding the volume of output and by uncertainty regarding its market price. Notwithstanding these uncertainties, formal insurance markets are typically absent. Instead, the form of social organization in agrarian communities is strongly influenced by the consideration of insuring against a subsistence crisis (J. C. Scott 1976; Posner 1980). The importance of risk in determining social arrangements such as family formation, migration, and remittances has been attested by empirical studies (Nelson 1976; Knowles and Anker 1981; Lucas and Stark 1985; Caldwell *et al.* 1986; Khandker *et al.* 1987; Rosenzweig 1988*a*, 1988*b*; Rosenzweig and Stark 1988; Shaw 1988; Stark and Lucas 1989) and there is little doubt that risk-sharing is also an important consideration in contract choice.

Another major characteristic of agricultural production in developing economies is the high enforcement costs of labourers that limit the growth of farm firms or plantations based on hired labour. As aptly described in a classic article by Brewster (1950), the scope for scale economies is rather limited in agriculture not only in developing economies but also in developed economies because, unlike industrial production dealing with lifeless and mobile materials, agricultural production is affected by seasonal time-sequences and location specificities. Because of this, agricultural production cannot be performed concurrently in one location and hence a greater division of labour, especially the specialization of managerial and supervisory functions separated from direct labour, tends to be unprofitable. The difficulty in monitoring hired labour becomes greater in more complex farming systems involving crop rotation and livestock: 'for multiple enterprise farms, family operators have the advantage. Increasing the number of enterprises so multiplies the number of on-the-spot supervisory management decisions per acre that the total acreage which a unit of management can oversee quickly approaches the acreage which an ordinary family can operate' (Brewster 1950: 71). In fact, large plantation operations are limited largely to monoculture.

This constraint of managerial ability and family labour on operational farm size is exacerbated by the danger of the misuse of draught animals and machines by non-family operators that results in capital loss. Therefore, 'a landless person with a family who owns animals and/or machines and possesses some managerial skill will find it more profitable to rent land than to hire out his endowments separately. Similarly, a large landowner will find it more profitable to rent out land than to manage a large operation because of scale diseconomies arising from the use of hired workers' (Binswanger and Rosenzweig 1986: 524). In other words, technological scale economies arising from the use of indivisible inputs such as managerial ability and animals/machines are counterbalanced by scale diseconomies from the use of hired labour. Therefore the nuclear family farm becomes the optimum form of organization except in the production of plantation crops that need close coordination with large-scale processing and marketing.

In high-income developed economies, however, sharply rising wage-rates in agriculture due to absorption of labour from the non-agricultural sectors have led to the development and diffusion of labour-saving agricultural technology. The increasing size of indivisible machine capital has enlarged the zone of increasing return to farm production. On the other hand, typically low and stagnant wage-rates in developing economies provide little incentive for large-scale farm mechanization to effect scale economies. Indeed, aggregate agricultural production function analysis based on intercountry cross-section data confirms that scale economies operate in developed countries whereas constant returns prevail in developing economies (Hayami and Ruttan 1985; Kawagoe *et al.* 1985). An extension of intercountry production function analysis reveals that the degrees of scale economies are positively associated with the level of tractorization (Lau and Yotopoulos 1989). Further, a comparison of cross-sectional estimates of the production function in Japanese agriculture shows that constant returns prevailed when wage-rates in Japan were still much lower than in North America and Western Europe and that scale economies began to emerge in the 1970s when large-scale mechanization was encouraged by higher wages (Hayami and Kawagoe 1989). At the micro-level, a large number of case-studies are available for developing economies, especially India, that observe inverse correlations between farm size and output per hectare mainly for food crops (Bhagwati and Chakravarty 1969; A. K. Sen 1975; Berry and Cline 1979). These inverse relations, however, are largely explained by differences in land quality and factor-market imperfections and therefore are not inconsistent with the hypothesis of constant returns in farm production (Verma and Bromley 1987; Bhalla and Roy 1988).

These cross-sectional estimates of constant returns should not be taken to indicate that the production function of individual farms is characterized by constant returns for the whole operational size range. It is more reasonable to assume that the individual farm production function has zones both of increasing and decreasing returns with the optimum size determined by the endowment of family-owned resources, especially the operator's managerial ability and family labour. Therefore, estimated constant returns are considered to reflect the fact that economic efficiency does not differ across farms of different optimum sizes. If this interpretation is valid, the production function at the individual farm level in developing economies can be legitimately considered to be characterized by constant returns so long as properly measured worker-effort (including management-effort) is included as an input variable (as specified in the model analysis in later chapters), since there is no reason to assume that production efficiency declines when the work-effort increases proportionally with other inputs. Indeed, the well-known study by Mundlak (1961) shows that scale economies estimated from production function analysis tend to disappear when the variable to represent management input is added to the conventional labour-input variable.

While the characteristics of agricultural production are expected to result in the dominance of family farms (either owner- or tenant-operated) in agrarian economies, the seasonal nature of agricultural production makes it difficult for peasant producers (both owner and tenant) to base their operations on family labour alone. At the peak of demand for labour, such as during harvesting, casual labour is commonly employed. Employment of casual labour is also common in weeding and rice transplanting; the result of this work as well as harvesting is relatively easy to measure so that the work of casual labourers can be supervised at relatively modest cost even under a time-rate contract or their work-incentives can be enhanced by applying a piece-rate contract (Roumasset and Uy 1980). Land-preparation work is also measurable in terms of area ploughed or harrowed. Yet hiring of casual labourers for land preparation is rare because of the danger of their harming draught animals and damaging machines through careless use. For the same reason, the renting of animals or machines alone is seldom practised. Instead, the hiring of a machine with operator or an animal and its owner for ploughing and harrowing is more common (Bliss and Stern 1982; Barker and Herdt 1985).

Water control, and fertilizer and chemical applications require care and judgement, and their contribution to output is difficult to measure. Furthermore, their labour requirements are spread thinly over a crop season in an unpredictable manner. These tasks are usually carried out by family labour or, exceptionally, by 'permanent labour' as a substitute for family labour (Hirashima 1978; Otsuka *et al*. 1993). The permanent labourer (alternatively called 'attached labourer') is usually employed for a crop season or a year and paid a fixed sum or a share of output or a combination of both, with various fringe benefits such as free board and lodging. Often, the contract is renewed over many years. Unlike the casual labourer who is employed for a specific task the scope of the permanent labourer's work is not clearly specified even though his major obligations such as herding animals are implicitly agreed upon. The strong personal attachment of the permanent labourer to his master and his long-run prospects of future benefits relative to the casual labourer to some extent discourage him from shirking and cheating. However, to the extent that the permanent labourer can claim less on residual farm profit than the tenant, his incentive to work is smaller. In agrarian economies outside the plantation sector, therefore, farms based on permanent labour are rather exceptional compared with the prevalence of tenant-operated farms.

Community and market

Markets for agricultural products in the Third World are often rudimentary and undeveloped. Because of poor transport and communication facilities, market information is scarce and costly to obtain and, as a result, markets tend to be segmented into the units of small local communities. Formal crop insurance is not feasible in such small communities, partly because in agricultural production it is difficult to distinguish the effects of natural hazards from those of mismanagement, and partly because production risks are highly correlated (Binswanger and Rosenzweig 1986; Binswanger and McIntire 1987). The small size of the market in each community precludes the possibility of using modern marketing practices aimed at reducing uncertainty about product quality. In fact, the marketable agricultural surplus of peasant producers is too small and variable to introduce grading and brand names. In the presence of quality uncertainty, regular transactions backed by a personal bond—'clientelization' as Geertz (1978) calls it—are much preferred to spot exchanges with strangers in the marketplace.

Since potential market size is usually too small for specialized agents to engage profitably in the marketing of goods and services separately, there is a strong tendency for many transactions to be interlinked through highly personalized relationships (Bardhan 1980; Hart 1986; Bell 1988). Interlinked contracts between the same parties can be good substitutes for specialized markets for labour, land, credit, and insurance. For example, as is common in share tenancy, a landlord does not merely receive a share rent for his contribution of land to the production process, but also he bears a part of production costs and advances credits for production and consumption purposes. Moreover, he often insures his tenant against unexpected hazards by reducing rent in the event of crop failure, giving gifts when one of the family is sick, and using his connections and influence to solve the tenant's troubles with outside authorities such as the police. The tenant reciprocates with his and his family's loyal service not only in farm production but also in social and political activities.

Such a relationship is commonly called by anthropologists and sociologists a patron–client relationship—'a special case of dyadic (two-person) ties involving a largely instrumental friendship in which an individual of higher socio-economic status (patron) uses his own influence and resources to provide protection and/or benefits for a person of lower status (client) who, for his part, reciprocates by offering general support and assistance, including personal services, to the patron' (J. C. Scott 1976: 8). In the patron–client relationship, exchanges are multi-stranded and the balance is supposed to be cleared in the long run. The multiple transactions between the same parties over many years and across goods and services, as manifested in the patron–client relationship, is a device to reduce transaction costs because much of the cost of collecting information and contract enforcement is common to all the transactions (Bardhan 1980; Hayami and Kikuchi 1982).

How effective such long-term interlinked contracts work as a device to reduce transaction costs depends on social structures. In the small community in agrarian economies, social interaction among people is intense. Everyone is watching everyone. One party's dishonest behaviour tends to be detected by neighbours in the community and conveyed quickly to another party by word of mouth. In such a community, both parties may be discouraged from behaving dishonestly given the severe cost of a loss of reputation in the event of discovery.

We do not consider that intensive social interactions in an agrarian community are always sufficient to enforce the terms of contract. According to the theory of social interactions by Becker (1974, 1976), a person will refrain from exercising opportunism to the extent that its expected cost exceeds the expected gain. Then one may be tempted to exercise opportunism if the probability of detection is small. For example, a landlord well experienced in farming may be able to enforce effectively the work of share-tenants or even permanent labourers as he is able to detect shirking from his field-observations as well as from neighbours' gossiping. However, his widow with little knowledge of farming may find it difficult to manage the farm even under a share contract and, therefore, will consider a fixed-rent tenancy the only feasible option. Thus, enforceability of contracts depends on the human capital endowments of individual contracting parties as well as community structures and agricultural environments, which determine the cost and benefit of committing a moral hazard and a breach of contract.

1.3 Plan of the Book

First, a basic one-period model embracing both land and labour contracts is developed in Chapters 2 and 3. Chapter 2 sets out the structure of the basic model, while Chapter 3 considers the optimum contract choice under alternative assumptions on the enforceability of the tenant's and labourer's work-effort and the risk-aversion of the contract parties. The basic model simplifies the general reward system in agency theory into a linear form that is consistent with observable contract modes in agrarian economies. With this simplification, the derivation of empirically testable propositions is greatly facilitated, while the basic structure of the agency model is maintained. Within the framework of this basic model, widely divergent perspectives on agrarian contracts are classified according to underlying assumptions. It will be shown that different theoretical perspectives stem from partial treatments of contract choice which unduly limit the optimizing behaviour of contract parties and the options of contract choice, as well as from different assumptions of contract enforceability.

The basic one-period model abstracts from the mechanism of contract enforcement and adopts the extreme assumption of either perfect or totally imperfect monitorability of the agent's work-effort. The principal's information on the agent's work-effort is expected to increase in the long run because the effects on production of uncertain factors such as weather tend to even out *ex post*. Moreover, over a longer period of time, increased options become available to penalize those who do not and reward those who do honour their contracts. In Chapter 4, we explicitly introduce the monitoring function of the agent's work-effort, which enables us to analyse the general case of imperfectly enforceable contracts. In this chapter, the permanent-labour contract is considered a close substitute for the long-term tenancy contract. An important conclusion is that the fixed-wage permanent-labour contract will be dominated by some form of tenancy contract so long as both are available options to the parties. In the light of recent developments in the theory of multiperiod contracts, we also re-examine the significance of wayward behaviour of workers/tenants assumed in various one-period models within the context of the long-term contract.

A major puzzle unexplained by existing contract theories is the stylized fact of share tenancy that output is almost universally shared between tenant and landlord on a 50:50 basis with no explicit fixed payments, despite obvious differences in the relative contributions of land and labour to production among different environments and technologies. The basic model abstracts from the fact that land and labour contracts are often interlinked with credit, insurance, and cost-sharing contracts. A review of existing models of interlinked contracts in Chapter 5 reveals that both the repayment of credit and the shared cost correspond to the fixed payment in the basic model and suggests that the puzzle of 50:50 sharing may be resolved if the discount factor over the crop season is taken into account.

Chapter 6 evaluates the empirical relevance of various land and labour contract models regarding contract choice and enforcement based on a global survey of past empirical studies. This review indicates that the incidence of fixed-wage permanent-labour contracts is commonly associated with socio-political regulations such as land-reform laws that prohibit land-tenancy contracts, and that the inefficiency associated with share tenancy is usually found in areas where the choice among various land-tenure forms is institutionally restricted. These findings have important policy implications in view of the fact that, under the influence of the Marshallian argument that share-tenancy contracts are inefficient, land-reform laws in many developing countries prohibit share tenancy (Ruttan 1964; Warriner 1969; Mangahas *et al.* 1976; Ladejinsky 1977; Koo 1982; Herring 1983; Prosterman and Riedinger 1987).

Chapters 7–9 report the results of our case-studies undertaken to develop more concrete and in-depth understanding of the mechanism of contract choice and enforcement. Chapter 7 deals with the case of a typical agrarian community in Java, Indonesia, in which social interactions are intense and where no institutional constraint operates to limit choice among various land and labour contracts. Chapter 8 analyses changes in the choice of contracts corresponding to a shift in an economy from one based on narrow community relations to one based on wide impersonal markets, taking the case of the informal minibus called a 'jeepney' in the Philippines. The results of these two case-studies are consistent with the general hypothesis that rural people in developing economies make efficient choices from a wide spectrum of contracts taking into account their own resource endowments and external conditions surrounding them. In contrast, Chapter 9 deals with the rice sector in the Philippines, where land-reform regulations strongly limit the scope of contractual options. Results of the analysis indicate that the artificial limitation of contract choice results in both inefficiency and inequity.

Finally, in Chapter 10 both theoretical and empirical findings are summarized in such a way as to point in the direction of establishing a general theory of contract choice in agrarian economies, and the possible implications of our agrarian perspective for economic organizations in general are discussed.

2

The Basic Model

IN this chapter we develop a basic one-period model of land-tenancy and labour-employment contracts based on the theory of the principal–agent relationship developed by Harris and Raviv (1978, 1979), Holmstrom (1979), and Shavell (1979), among others, while drawing on the assumptions common to several representative theories of share tenancy advanced by Cheung (1969), Bardhan and Srinivasan (1971), and Stiglitz (1974).

An agency problem typically arises when one party, designated as agent (e.g. the tenant), acts on behalf of another party, designated as principal (e.g. the landlord), who lacks perfect information on the agent's action. There is conflict of interest between the contracting parties where the agent's work-effort contributes to increased output and the principal's income positively depends on output: the principal wants the agent to work harder, whereas work-effort generates disutility to the agent. Because of imperfect information on the agent's work-effort, the principal cannot specify and enforce the desired level of effort. Moreover, because of the unmeasurable effect on production of uncertain factors, he cannot verify whether low output is due to the agent's shirking or to unfavourable external factors. Thus, the agent is assumed to choose his action so as to maximize his (expected) utility given the structure of his reward function, while the principal selects the reward function that maximizes his own (expected) utility. In addition, in order to attract the agent, the principal must guarantee him utility no less than his reservation (or opportunity) utility (i.e. the agent's best available utility from a contract with some other principal or from some other activity).

If the principal can observe and enforce the agent's work-effort, there is no agency problem; the principal simply specifies the desired level of effort in the contract and forces the agent to observe it. Thus, in the basic model we focus on the problem of enforcing the work-effort of tenants and permanent labourers, who are mainly engaged in care-intensive activities such as water control, fertilizer and chemical application, and animal care which are not amenable to easy supervision. We exclude casual labour from our consideration here because casual labourers are mainly engaged in such activities as weeding, transplanting, and harvesting, which can be easily monitored, as we argued in Section 1.2. In reality, the casual-labour market is active and virtually all farming households hire casual labourers at peak seasons. Yet the inclusion of casual labour does not affect the substance of the model so long as its work-effort is monitorable and hence enforceable.[1] Perfect enforceability of casual labour will be assumed in Chapter 4 which deals with long-term contracts.

An agency problem also arises when the principal is not familiar with the characteristics (e.g. ability) of the agent, which gives rise to the so-called adverse selection problem. The principal is willing to offer high remuneration only to the agent with suitable ability, because this contributes to increased output and, hence, the principal's income, whereas the agent attempts to choose the contract which maximizes his utility given his ability. The problem for the principal here is to devise a set of contracts which will induce the desired self-selection of contracts by agents. The general presumption of agricultural tenancy and labour-contract literature, however, is that the abilities of workers are fairy well known to landlords within a closed traditional village economy (see e.g. Bardhan 1984: 97; Bell 1988). We therefore focus our analysis in this chapter on the issue of work-incentives and risk-sharing, leaving the analysis of self-selection to the next chapter.

While our model follows the basic structure of the agency model, we confine the agent's reward function to the class of functions linear in output: this assumption is common to the theoretical literature on the tenancy contract (see, among others, Stiglitz 1974; Mitra 1983; Braverman and Stiglitz 1982, 1986*a*; and Otsuka and Hayami 1988). This simplifying assumption greatly facilitates the derivation of optimum conditions and the intuitive understanding of the resulting optimum contract.[2]

2.1 Specification of Functional Forms

We assume that there are N homogeneous landless farm-workers and a single cultivating landlord who owns H hectares of land. The 'landless farm-worker' here refers either to a tenant or to a

The Basic Model

permanent labourer. We consider a one-season contract here, even though in practice the tenancy and permanent-labour contracts usually last for an extended number of periods. The production function for the landless worker is defined as

$$Q = \theta F(L,H), \tag{2.1}$$

where Q is output per worker; θ represents the state of nature, which is influenced by the vagaries of the weather; L and H are the work-effort of the worker and land input per worker, respectively; and function F is a concave, linear homogeneous function with the usual properties of positive but decreasing marginal products, i.e. $F_1, F_2 > 0$, and $F_{11}, F_{22} < 0$. Following Stiglitz (1974), θ is treated here as a multiplicative factor distributed with $E\theta = 1$ and finite variance. Input decisions are assumed to be made before θ is realized.[3] The unit price of output is exogenously given and normalized to unity. The work-effort (L) is not the conventional variable of labour input as measured by labour days but includes the worker's conscientious effort to apply his labour appropriately for various farm tasks under diverse ecological conditions.

In addition to its analytical convenience, the assumption of constant returns is adopted in view of the nature of the agricultural production function in developing economies as discussed in Section 1.2. If the endowment of the farm-operator's management ability and family labour that are the source of the work-effort are the single indivisible input to create scale economies or diseconomies as argued by Binswanger and Rosenzweig (1986), there is no reason to assume that production efficiency increases or decreases when the worker's effort (L) increases proportionally with the other input (H). However, its replacement by the assumption of decreasing returns will not affect major conclusions except for the determination of land input per worker (Newbery 1975a).[4]

Following Bardhan and Srinivasan (1971), a landlord is assumed to farm a part of the land (h) himself with his own work-effort (ℓ). Although the production functions of landlord and worker are assumed to be the same, the landlord's are expressed by lower-case letters so as to be distinguishable from the worker's:

$$q = \theta f(\ell, h). \tag{2.2}$$

Following Cheung (1969), it is assumed that the landlord owns a certain amount of land ($\bar{H}$) and rents out equal amounts of land to

N homogeneous workers, and hence $h = \bar{H} - NH$. Therefore, equation 2.2. can be rewritten as:

$$q = \theta f(\ell, \bar{H} - NH). \tag{2.3}$$

Again, following Stiglitz (1974), the return to a worker (Y) is assumed to be expressed by the following linear function:

$$Y = \alpha Q + \beta, \tag{2.4}$$

where Y is his income, α is a parameter representing the output-sharing rate, β is a parameter representing the fixed payment which corresponds to a fixed-wage component if $\beta > 0$ and to a fixed-rent component if $\beta < 0$. Typical forms of land and labour contracts can be expressed by different combinations of α and β as follows:

Fixed-wage labour contract: $\alpha = 0, \beta > 0$;
'Pure' share-tenancy contract: $0 < \alpha < 1, \beta = 0$;   (2.5)
Fixed-rent tenancy contract: $\alpha = 1, \beta < 0$.

The 'pure' share-tenancy contract is the common form of sharecropping tenancy. Formally it is the same as the piece-rate labour contract (Stiglitz 1975) commonly adopted in the employment of casual labourers in harvesting and threshing activities (Roumasset and Uy 1981; Hayami and Kikuchi 1982).[5] There is, however, the difference between the two in practice: the share contract when applied to a specific task (such as harvesting) is called a piece-rate labour-employment contract, whereas it is called a share-tenancy contract where a worker is assigned overall tasks in the farm including farm management decisions for a long-term period (at least one crop season).

There are also cases where the share contract applies to the employment of permanent labourers. In India, an output-sharing arrangement was traditionally prevalent with permanent-labour contracts (Sanghavi 1969; K. Bardhan 1977). Even in recent years the incidence of permanent-labour contracts, in which output was shared (sometimes in a 50:50 ratio) between employers and labourers, is reported by a large number of village studies (Thorner and Thorner 1962; Breman 1974; Bhalla 1976; K. Bardhan 1977; Bell 1977; Bardhan and Rudra 1981). Richards (1979) also observes output-sharing arrangements under permanent-labour contracts in Egypt and Chile in the nineteenth and early twentieth centuries.

The Basic Model

Thus, the distinction between the share-tenancy contract and the permanent-labour contract with an output-sharing arrangement is not always easy to draw because the major difference lies in the degree to which landlords and workers are involved in management and this is sometimes difficult to ascertain. In our theoretical model, the share contract described by equation 2.4 or 2.5 subsumes both output-sharing tenancy and labour contracts.

Usually, under the share contract no explicit fixed-crop or cash payment is made between the worker and the landlord. This is perhaps the major reason why share-tenancy models before Stiglitz (1974) assumed $\beta = 0$.[6] One might expect to observe a great variety of contracts with different values of α and β. Output-sharing rate is, however, almost universally 50 per cent under share tenancy, notably in eighteenth- and nineteenth-century France (H. Higgs 1894), in the *post-bellum* South in the US (Reid 1979b), in post-war California (Wells 1981), and in many contemporary developing countries (e.g. C. H. H. Rao 1971; Zaman 1973; Pal 1975; Mangahas et al. 1976; Jabbar 1977; Bardhan and Rudra 1980; Bell and Srinivasan 1985a; Nabi 1986). Although it is not entirely unaffected by technology, land quality, and cost-sharing arrangements of purchased inputs (Vyas 1970; Shlomowitz 1979; Bardhan and Rudra 1980; Khasnabis and Chakravarty 1982; Roumasset and James 1979; Fujimoto 1983; Roumasset 1984; Morooka and Hayami 1989), the fact remains that it is predominantly one-half. When it is not one-half, it is almost always two-thirds. It is interesting to note that the French and Italian words for share tenancy, *métayage* and *mezzandria* respectively, literally mean splitting in half. This stylized fact remains the 'major puzzle' of share tenancy (Stiglitz 1988).

The resolution of the puzzle that so many contracts fall into the special cases described by equation 2.5 is taken up in Chapter 5 which deals with the interlinking of tenancy contracts with credit and cost-sharing contracts. For the time being, we simply regard the share contract as being characterized by α greater than zero but less than unity with no regard to the value of β.

The income of the landlord (y) is given by

$$y = N[(1 - \alpha) Q - \beta] + q. \tag{2.6}$$

The utility of the landless worker (U) depends on his income and effort:

$$U = U(Y,L), \qquad (2.7)$$

where it is assumed that $U_1 > 0$, $U_2 < 0$, $U_{11} \leq 0$, $U_{22} \leq 0$, and $U_{12} \leq 0$. The first two inequalities imply that the worker's utility increases with Y but decreases with L, whereas the third condition ($U_{11} \leq 0$) implies that he is either risk-averse or neutral. The last two conditions state that the marginal disutility of effort increases with effort ($U_{22} \leq 0$) and with income ($U_{12} \leq 0$), which ensures that 'leisure' is a normal good. Thus, when β decreases, the worker will increase his effort, which results in higher output. It is assumed that he maximizes his expected utility, $EU(Y, L)$.

The worker will not enter into the contract unless it provides him with utility at least equal to his reservation utility (V). Therefore, the following reservation-utility constraint must hold:

$$EU(Y,L) \geq V, \qquad (2.8)$$

where V is assumed to be fixed, implying that this is a partial equilibrium analysis.[7] The utility function of the landlord is similarly defined as

$$Eu(y,\ell), \qquad (2.9)$$

where u is assumed to be concave with the same property as for U.

2.2 The Formulation of the Optimization Problem

The contract must be made based on variables observed by both the landlord and the landless worker. Otherwise, conflict will arise *ex post*. In formulating the reward function 2.4 which depends on output, it is assumed that output is observable to both contracting parties. It is also assumed that the fixed payment at the time of harvest is enforceable so that the defaults of fixed-rent payment by the worker when $\beta < 0$ and of fixed-wage payment by the landlord when $\beta > 0$ do not occur *ex post*. To be strict, as Holmstrom (1981, 1983) and Bull (1983, 1987) argue, in order that the contract is perfectly enforceable, its terms and conditions must be verifiable not only to the contracting parties but also to a third party, e.g. a court, which can penalize those who violate the terms of contract agreed upon. Breach of contract (e.g. default of rent payment) is a real possibility in the case of a one-period contract because of the

The Basic Model

absence of future penalties (e.g. tenant eviction with the high cost of recontracting). We will discuss this issue in more detail when we analyse the long-term contract in Chapter 4. In the basic model, we assume that a sufficiently large penalty, either pecuniary or non-pecuniary, will be imposed on breaches of agreement on rent and wage payments so that rational parties will honour these agreements.

We also assume that there are many landless workers in the economy so that the reservation utility is exogenously given to each contracting party. If there are a relatively small number of workers, V will not be exogenously given to the contract parties and a bargaining problem will arise. We will examine the model of Bell and Zusman (1976) in Chapter 3, which applies the Nash bargaining approach to the solution of share-tenancy equilibrium.[8]

Provided the worker receives a utility of V, the contractual parameters, α and β, are determined by the landlord to maximize his utility. We also assume that the amount of land per worker, H, is determined by the landlord because unlike labour the use of land is easily observable.[9] Note that the determination of the contractual parameters, α and β, and H by the landlord does not necessarily imply that he has a monopoly power. He would be a monopolist if the worker had no alternative job opportunities and the reservation utility reflected his utility when unemployed. On the other hand, he is a competitive landlord if the worker's reservation utility represents the level of utility he can acquire in a competitive labour market including tenancy arrangements with other landlords. Formally the implications of the model are the same whether the market is competitive or monopolistic because V is constant in these cases (Braverman and Stiglitz 1982).

Nevertheless, Cheung's (1969) pure share-tenancy model has often been labelled as a monopoly model because of the assumption that α and H are determined by the landlord (Bardhan and Srinivasan 1971; Koo 1973, 1977; Currie 1981; Jaynes 1982; Quibria and Rashid 1984). Bardhan (1977, 1979a) adopts the assumption that the landlord determines H because he has monopolistic power, while Bardhan and Srinivasan (1971), Mangahas (1975), Mangahas et al. (1976), Pant (1983), Alston et al. (1984), Quibria and Rashid (1986), and J. M. Rao (1987) assume that H is determined by the worker because they regard α as a price-like parameter competitively determined in the market.[10] These assumptions are invalid. In market transactions of ordinary goods and services, the price is the only term of contract. But in the contract, it is the combination of the variables that determines its terms. If the landlord must guarantee to the worker the reservation utility competitively determined in the market, the model is a competitive one. It should be clearly recognized that the contracting parties in the competitive contract model are 'utility-takers' analogous to 'price-takers' in the competitive market model of ordinary goods and services.

A major controversy in the tenancy and labour-contract literature centres on the enforceability of the worker's work-effort. The farming activities of tenants and permanent labourers involve land preparation, fertilizer application, and even the supervision of casual labourers. These activities require considerable care and judgement. Thus, worker's effort refers not only to simple manual labour but also to decision-making activities. The traditional and still dominant view is that L is unenforceable by the landlord in part because of the spatial nature of agricultural production and in part because of the landlord's difficulty in identifying L from observations on Q and H in the presence of production uncertainty represented by θ. If the landlord is familiar with farming technology, he should be able to ascertain the effect of weather on θ accurately *ex post* and estimate L from the production function reasonably well (Newbery 1975a; Kotwal 1985). However, the realized value of θ specific to a certain area would not be verifiable to a third party and, therefore, L still would not be enforceable by the landlord. In the tenancy literature, L is widely assumed to be totally unenforceable.

So L becomes the worker's choice variable. The worker works effectively when $\alpha = 1$; when $\alpha < 1$, he is said to shirk and a moral-hazard problem arises. In this case the worker maximizes EU with respect to L, implying a first-order condition:[11]

$$EU_1 \; \theta \alpha F_1 + EU_2 \leq 0. \tag{2.10}$$

This suggests that under the fixed-wage contract ($\alpha = 0$), the worker has no incentive to work at all. Hence the inequality holds in equation 2.10 and he will choose $L = 0$, which is obviously inefficient. Resource allocation under the share contract ($1 > \alpha > 0$) is not first-best efficient, since the expected marginal product of effort (F_1) is not equated with the marginal rate of substitution between effort and the expected output ($-EU_2/EU_1\theta$). Therefore,

The Basic Model

because of the problem of monitoring the worker's effort, the share contract does not achieve first-best efficiency.

The traditional Marshallian thesis of share tenancy assumes that production is certain and argues from equation 2.10 that share tenancy is inherently inefficient. Once certainty is assumed, however, the problem of enforcing L essentially disappears because L can be readily inferred from Q and H. Therefore, the assumption of certainty ($\theta = 1$) should be replaced by the risk-neutrality of landlord and worker ($u_{11} = U_{11} = 0$) under uncertainty, even though the implications of the models are formally identical under the two sets of assumptions.

Major advocates of the Marshallian thesis are, listed chronologically: Schickele (1941), Heady (1947), Castle (1952), Drake (1952), Issawi (1957), Georgescu-Roegen (1960), Bhagwati (1966), A. K. Sen (1966), Adams and Rask (1968), J. T. Scott (1970), Bardhan and Srinivasan (1971), Koo (1973), Mangahas (1975), Mazumdar (1975), C. H. H. Rao (1975), A. K. Sen (1975), Bell and Zusman (1976), Mangahas et al. (1976), Bardhan (1977, 1979a, 1984), Bell (1977), Hurwicz and Shapiro (1978), Lucas (1979), Braverman and Srinivasan (1981), Pant (1983), Alston et al. (1984), Eswaran and Kotwal (1985b), Quibria and Rashid (1986), J. M. Rao (1987), and Shetty (1988). With the exception of Bell and Zusman, Hurwicz and Shapiro, Eswaran and Kotwal, and Shetty, the Marshallian models assume that α is exogenously given. Bell and Zusman as well as Hurwicz and Shapiro assume $\beta = 0$, thereby excluding the choice of fixed-rent contract. These assumptions unduly limit the options of the contract and are inconsistent with the rational behaviour of the landlord. We will review critically the Bell–Zusman model in Chapter 3 and the Eswaran–Kotwal model in Chapter 4.

More recent tenancy models derive equation 2.10 under the assumptions of uncertainty, risk-aversion of the worker, and the endogeneity of contractual parameters (Stiglitz 1974; Hebert 1978; Newbery and Stiglitz 1979; Braverman and Stiglitz 1982, 1986a, 1986b; Mitra 1983; Gangopadhyay and Sengupta 1986; Bell 1988).[12] Under the assumptions of these 'neo-Marshallian' models equation 2.10 implies a reaction function for the worker's effort of the form:

$$L = L(\alpha, \beta, H), \qquad (2.11)$$

in which work-incentives are affected by α, β, and H.[13] The landlord's problem is to maximize his expected utility with respect to α, β, and H, subject to the worker's reaction function of effort (equation 2.11) and to his reservation-utility constraint (equation 2.8), i.e.

$$\max_{\{\alpha,\beta,H,N,\ell\}} Eu[N\{(1-\alpha)\,\theta F(L(\alpha,\beta,H),H) - \beta\} + \theta f(\ell, \bar{H} - NH),]$$

$$\text{s.t.} \quad EU[\alpha\theta F(L(\alpha,\beta,H),H) + \beta, L(\alpha,\beta,H)] = V. \quad (2.12)$$

Note that according to this characterization of the problem, the landlord can affect the worker's level of effort only by manipulating α, β, and H.

By contrast, Johnson (1950) and Cheung (1969) assume that the landlord can observe and enforce the worker's effort perfectly. Earlier, Marshall (1890: 563) also considered the solution under the enforceability assumption, although his argument was not correctly appreciated until Bliss and Stern (1982). If L is enforceable, the landlord determines not only α, β, and H but also L so as to maximize his expected utility, subject to the reservation-utility constraint, $EU = V$, alone, i.e.

$$\max_{\{\alpha,\beta,H,L,N,\ell\}} Eu[N\{(1-\alpha)\,\theta\, F(L,H) - \beta\} + \theta f(\ell, H - NH), \ell]$$

$$\text{s.t.} \quad EU[\alpha\theta F(L,H) + \beta, L] = V. \quad (2.13)$$

Since the information is perfect, the first-best resource allocation and the optimum risk-sharing are attained (Stiglitz 1974; Otsuka and Hayami 1988).[14] However, the assumption that the worker's effort is perfectly enforceable is widely criticized as unrealistic.

Confusion still exists in the literature on the implications of the enforceability assumption for work-incentives under different contracts. Bell and Braverman (1980) adopt the peculiar assumption that the worker's effort is enforceable under the fixed-wage contract but unenforceable under the share contract. They assert that the fixed-wage contract always dominates the share contract because the former is efficient and the latter is inefficient. Such an assumption is totally untenable because the worker has no incentive to work under the fixed-wage contract, so that the cost to the landlord of enforcing the work-effort of the wage labourer ought to be higher than that of the share tenant. Nevertheless, such leading economists on share tenancy as Datta and Nugent (1985) and Bardhan (1985) have uncritically accepted Bell and Braverman's conclusion even recently.

Independent of the effort-reaction function shown in equation 2.1 (and hence the enforceability of the worker's work-effort) and the worker's reservation constraint, the following optimization rule can be derived from the choice of N:

$$Eu_1[(1 - \alpha)Q - \beta] - Eu_1\theta Hf_2 \leq 0, \qquad (2.14)$$

which states that the landlord's gain in marginal utility associated with an additional contract must not be smaller than his forgone marginal utility associated with the use of H on his owner-cultivated farm. If the former is always smaller than the latter, the landlord does not offer any contract and becomes a pure owner-cultivator. This occurs when the worker's cultivation is relatively inefficient, so that the landlord's income from the contract is small compared with the expected contribution of land to output on his own farm (Hf_2). Thus the dominance of small family-farms with the use of own-family labour supplemented by the employment of casual labour, observed in Chapter 1, may be explained by the inefficiency of the worker's cultivation relative to the landowner's. Another possible explanation is that the landlord's ownership of land is small relative to the endowment of his (and his family's) labour, so that the expected marginal product of land (f_2) with the use of the whole area is relatively large. This can be the case, even if the worker's cultivation is efficient. In that case, the dominance of owner cultivation will be explained by a relatively equal distibution of land ownership rather than the inefficiency of the worker's cultivation. On the other hand, if the landlord is not adept at farming, ℓ tends to be small and hence f_2 will also be small. In such a case, he may not undertake any owner-cultivation. In this way, the relative efficiency of owner's and worker's cultivation and the relative endowment of labour and land resources among households have profound implications for the organization of production. We will review the empirical evidence on the relative efficiency of owner- and tenant-cultivation in Chapter 6.

In the next chapter we will illustrate the optimum contracts in the basic one-period model under different assumptions regarding production uncertainty (or more correctly risk-attitudes of the contracting parties) and the enforceability of the worker's effort. We also consider the interior optimum by assuming that equality holds in equation 2.14. We will review the various extensions of the basic model in Chapters 4 and 5.

Notes to Chapter 2

1. See Otsuka *et al.* (1992) for the model which explicitly incorporates casual labour contracts into the basic model.
2. In recent surveys of the agency literature, Hart and Holmstrom (1987) as well as Arrow (1985) raise the practical question of whether the implementation of more general, non-linear payment schemes is feasible. In reality, the reward function in land tenure is only approximately linear (Ho 1976).
3. It is implicitly assumed here that the realized value of θ will be observed by the worker but not by the landlord after the contract starts. As is well known in the literature, once the individuals have agreed on an optimum contract, there cannot be any mutually beneficial new agreements after they receive their private information (Migrom and Stokey 1982; Holmstrom and Myerson 1983). Thus, observation of θ by the worker will not alter the initial contract agreed upon.
4. Once the assumption of decreasing returns is introduced, the so-called 'agent–agent problem' specific to team production arises (Holmstrom 1982; Carmichael 1983; Green and Stokey 1983). We do not discuss this problem in this chapter because it has not been dealt with in the past literature of agrarian contracts as it is not so relevant to agrarian economies. The assumption of constant returns, in effect, precludes consideration of this problem from our analysis.
5. Share or piece-rate contracts are common in other industries as well, e.g. see Platteau and Abraham (1987) for the case-study of fishing contracts in India, Otsuka *et al.* (1986) or Ch. 8 for the jeepney (minibus for public transportation in the Philippines), and Otsuka and Murakami (1989) for the taxi in Japan.
6. An exception is Marshall (1890), as mentioned in Ch. 3.
7. Mitra (1983) has proved that the reservation-utility constraint holds with an equality in equilibrium if $U_{12} \leq 0$. This is because, while a decrease in β always reduces EU, it increases y not only directly but also indirectly through increased output induced by the decrease in β. Therefore, the landlord, if rational, pushes the worker's utility to the point of equality with V in equilibrium. Thus, our basic model assumes $EU = V$. This also holds true if, instead of β, h is manipulated by the

The Basic Model

landlord, so long as $U_{12} < 0$ (Braverman and Srinivasan 1981). The case with $EU > V$ will be dealt with in Ch. 4, where we examine the theory of the permanent-labour contract.

8. Even if V is given to each contracting party, V will change as the aggregate demand for workers changes with changes in contractual terms if an agrarian economy as a whole faces an inelastic supply of workers. In such a general equilibrium analysis, certain conclusions on the distributional consequences of contract choice can be different from those delivered by partial equilibrium models because of changes in V. See Braverman and Stiglitz (1982) and Bell (1988) for the general equilibrium implications of contract choice.
9. It is, however, possible to let the tenant determine H if the rent of R per unit of area (i.e. $\beta = -RH$) is determined by the landlord. In this case, the landlord can perfectly control H by manipulating R. See Newbery (1974, 1975a), Reid (1976b), and Jaynes (1982).
10. In the model of Bardhan and Srinivasan, the marginal product of land drops to zero in 'equilibrium', as can be ascertained from the maximization of worker's utility with respect to H, which results in $EU\theta\alpha F_2 = 0$. Obviously, such an equilibrium cannot exist in a land-scarce economy, as was pointed out earlier by Johnson (1950) and later by Newbery (1974, 1975a).
11. Our analysis is confined to the case where the first-order approach is appropriate. For the validity of such approach for the global optimum, see Jewitt (1988).
12. Stiglitz, Mitra, Gangopadhyay, and Sengupta, however, also consider the case where L is enforceable by the landlord.
13. Stochastic variable θ is not included in equation 2.11 because by assumption the worker commits the effort-level L before θ is realized.
14. First-best optimum in the context of the agency model does not necessarily refer to the true social optimum, since the model is designed to examine the implications of the lack of complete, state-contingent markets (Hart and Holmstrom 1987; Levinthal 1988). In the context of our model, it is implicitly assumed that crop-insurance markets are absent. See Binswanger and Rosenzweig (1986) and Binswanger and McIntire (1987) for further details of insurance markets absent in less developed agrarian economies.

3

Optimum Contract Choice under Alternative Assumptions

THIS chapter attempts to identify within the framework of our basic model the conditions under which the various contracts are chosen as a result of the optimizing behaviour of both landlord and landless farm-workers. Mathematical expressions in the text are limited to major conclusions and their derivations are shown in the Appendix to this chapter.

3.1 Optimum Contract Choice under 'Certainty'

A large number of tenancy models have been developed without assuming uncertainty or risk-aversion. As discussed in the previous chapter, the assumption of certainty should not be literally interpreted as $\theta = 1$ but understood as the risk-neutrality of the landlord and the worker under uncertainty because the enforcement problem of L does not exist under certainty. The existing share-tenancy models often assume the existence of a competitive labour market outside the farming sector. Therefore, it is assumed in this section that the worker expends L' units of work-effort at a competitive wage-rate (W) besides expending L on the farm, as assumed in previous models. Accordingly, equation 2.8 is reformulated as

$$U(\alpha Q + \beta + WL', L + L') = V. \tag{3.1}$$

This reformulation, however, does not affect the substance of the analysis.

Problems of the Marshallian thesis

The so-called Marshallian thesis of inefficiency in share contract has assumed an unenforceable contract in that the worker determines the level of L. (The 'Marshallian' thesis is not necessarily the same as Marshall's, as will be discussed later.) Maximization of U in equation 3.1 by the worker with respect to L and L' results in

$$U_1\alpha F_1 + U_2 = 0 \tag{3.2}$$
$$U_1 W + U_2 = 0. \tag{3.3}$$

From the above equations we obtain

$$\alpha F_1 = W. \tag{3.4}$$

This is the most familiar equation of Marshallian inefficiency in that only a fraction of marginal product of effort is equated with the competitive market wage-rate. It is clear that the worker shirks so long as $\alpha < 1$.

The optimization problem of the landlord can be solved by maximizing his utility subject to equations 3.1, 3.2, and 3.3. From the first-order conditions with respect to α and β the following important relation can be derived (see the Appendix to this chapter):

$$(1 - \alpha)(F_1/F_{11}) = 0, \tag{3.5}$$

which means that in equilibrium the fixed-rent contract ($\alpha = 1$) is adopted (Warr 1978; Currie 1981). That is, if the enforcement of the worker's work-effort alone is the inherent problem of the contract, the landlord will prefer the fixed-rent contract to share and fixed-wage contracts because of its non-distortionary incentive effect. This result is consistent with the well-known theorem of the theory of agency that equilibrium is characterized by the payment of a fixed fee from the agent to the principal if the agent is risk-neutral (Harris and Raviv 1978, 1979; Holmstrom 1979; Shavell 1979). It can easily be shown that $F_1 = W$, $F_2 = f_2$, $-\beta = HF_2$ for $\alpha = 1$; these are the conditions of Pareto optimality. It should be clear that the equilibrium of Marshallian inefficiency under share contract is derived from the optimization of the worker alone without due consideration of the landlord's optimization behaviour with respect to the choice of contract. It can easily be shown that if the landlord is allowed to choose α and β, the share contract cannot exist in recent Marshallian models (Mazumdar 1975; Bell and Zusman 1976; Bardhan 1977, 1979a; Lucas 1979; and Alston, et al. 1984), even though additional complexities. such as the existence of unemployment in the external labour market and the cost of supervising fixed-wage workers, are introduced in these models. The bargaining model of Bell and Zusman will be further examined later in this chapter.

Efficiency hypothesis under 'certainty'

A counterthesis to the Marshallian theory was developed by Cheung (1969) based on the optimization of the landlord. He showed that the share contract can achieve Pareto optimality under the condition of zero enforcement cost. His model assumes that in equilibrium the excess 'profit' of the worker (Y') is zero so that the excess supply of workers who seek share contracts does not emerge, that is

$$Y' = \alpha F(L,H) = \beta - WL = 0. \tag{3.6}$$

In the original Cheung model, β is assumed to be zero, but here it is assumed more generally that β is also determined endogenously. In this generalized Cheung model the optimization of the landlord amounts to the maximization of the landlord's utility with respect to α, β, N, H, ℓ, and L subject to the constraint of equation 3.6. This procedure was criticized by Bardhan and Srinivasan (1971) on the ground that the Cheung model does not incorporate the worker's optimizing behaviour. Their criticism misses the point because equation 3.6 has the same implication for the equilibrium as the combination of equations 3.1 and 3.3, in which the worker maximizes his utility with respect to his wage-labour (L').[1]

The landlord's optimization produces Pareto-optimum conditions as follows:

$$F_1 - W = 0; \tag{3.7}$$
$$f_2 - F_2 = 0; \tag{3.8}$$
$$(F - WL) - Hf_2 = 0; \tag{3.9}$$
$$u_1 f_1 + u_2 = 0. \tag{3.10}$$

Due to the assumption of a linear homogenous production function, equations 3.7, 3.8, and 3.9 are not mutually independent. Thus, the above equations determine only the optimum levels of NH, L/H, and ℓ. By inserting the optimum effort/land ratio into equation 3.6, a linear combination of α and β is given that generates the Pareto optimality for a given land area. The results imply that the Coase theorem also applies to the tenancy contract if enforcement cost is zero (Coase 1960; Hsiao 1975; Roumasset 1979).

It is important to emphasize that in this generalized model α and β are not uniquely determined, but an infinite number of their combinations exist in equilibrium. No single combination is preferred to any other, and, therefore, there is no positive reason for the choice of share contract. This point was not properly recognized by Cheung because he assumed that $\beta = 0$.

Cheung criticized the Marshallian thesis on the ground that the inefficiency of the share contract is derived from the irrelevant assumption of α being fixed. But he was not aware of the fact that Marshall (1890) himself recognized the possibility of achieving an efficient solution within the framework of the share contract by adjusting β for a given α, as pointed out by Bliss and Stern (1982).

Diagrammatic presentation

In order to facilitate understanding of the implications of the Marshallian and the Cheungian models, the discussion in this section will be restated in terms of Fig. 3.1, which has traditionally been used for illustrating the Marshallian inefficiency. For the

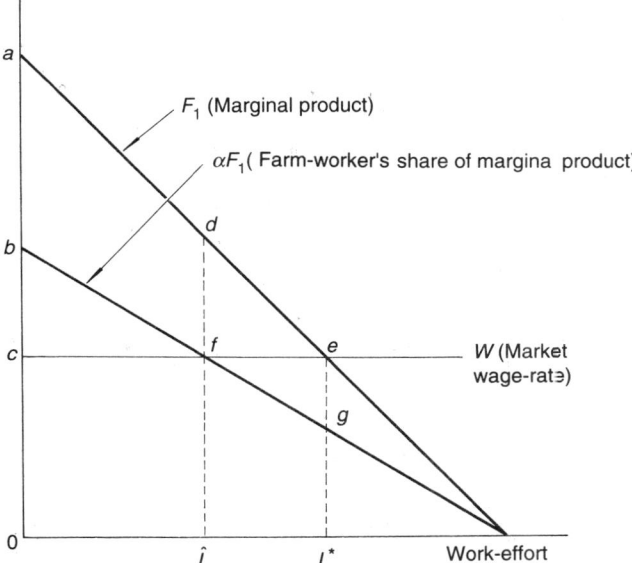

FIG. 3.1 Model of contract choice under 'certainty'

sake of simplicity, only the problem of determining the optimum level of L will be discussed, while assuming 'pure' share contracts ($\beta = 0$) alone as an alternative to fixed-rent tenancy and fixed-wage labour contracts.

In Fig. 3.1, under the Marshallian assumption of unenforceable contract for a given amount of land (H) and a given share rate (α), the worker's optimization is established at f, at which his share of the marginal product of work-effort (αF_1) is equated to the competitive market wage-rate. Work-effort under the share contract ($\hat{L}$) is lower than the Pareto-optimum work-effort (L^*), which corresponds to the equality between the marginal product of effort (F_1) and the market wage-rate (W) to be established at point e under fixed-rent tenancy and owner-farming. However, as was questioned first by Johnson (1950), the equilibrium cannot be established at f. As the rent that the landlord receives under the share contract (area $adfb$) is smaller than that under the fixed-rent tenancy (area aec), no landlord will accept the share contract. At the same time, the worker's share income from farming (area $bf\hat{L}O$) is higher than the income that he can earn by the equivalent amount of labour in fixed-wage employment in the competitive market (area $cf\hat{L}O$) by area bfc. Thus, the reservation-utility constraint holds with strict inequality. The excess profit by area bfc will give rise to an excess supply of workers seeking share contracts because all will prefer them to other forms of contract.

The excess supply arises in the solution of the Marshallian model because it neglects the landlord's optimizing behaviour. If the landlord's optimization is considered, he will not accept a share rent lower than the fixed rent. He will try to eliminate the worker's excess profit by requesting him to pay a fixed sum equivalent to area bfc ($\beta < 0$), and he will also try to raise α to induce the worker to increase his work-effort. In equilibrium the system will converge to the fixed-rent tenancy contract, in so far as the terms of contract of the worker's effort are difficult to enforce. The Marshallian solution can exist in equilibrium only when the landlord is an altruist who, as a benevolent patron, supports the income of poor clients at the expense of his income.

Under the Cheungian assumption of enforceable share contracts, the worker's effort is contractually specified at an optimum level (L^*) for a given H. Output-sharing rate α is adjusted by the landlord so that the share rent equals the fixed rent, that is, area bfc = area feg. However, the same equilibrium can also be achieved for a given α through the manipulation of β, as suggested by Marshall himself.[2]

In either case the resource allocations and the income distributions under share contract are the same as under the other forms of land-tenure if the contract is enforceable. In contrast, if share contract exists under the Marshallian assumptions, the work-effort and therefore the yield levels, as well as the land rent, will be lower than under other forms of land tenure, although the worker's income from farming under share contract (area $bf\hat{L}O$) can be larger or smaller than under fixed-rent tenancy (area $ceL*0$). The implications of these two different assumptions about work-enforcement under 'certainty' are summarized in the first two rows in Table 3.1.

TABLE 3.1. *Relative magnitudes of output, input, land rent, and tenant income from farming among three land-tenure classes, implied from different assumptions about contract enforcement*

Assumption	Yield (and input) per ha.	Rent per ha.	Workers' income from farming per ha.
Under certainty:			
Enforceable	S = R = OF	S = R	S = R
Unenforceable	S < R = OF	S < R	S ≷ R
Under uncertainty:			
Enforceable	S = R = OF	S ≥ R	S ≤ R
Unenforceable			
$\sigma < 1$	S < R = OF	S > R	S < R
$\sigma = 1$	S = R = OF	S ≥ R	S ≤ R
$\sigma > 1$	S > R = OF	S ≤ R	S ≷ R

Note: S = share contract; R = fixed-rent contract; OF = owner-farming.

3.2 Can Share Contract Exist under 'Certainty'?

In the previous section it was shown that under 'certainty', or more correctly in the absence of risk-aversion under uncertainty, share contract cannot exist under the assumption of unenforceable contracts and there is no definite reason for share contract to exist in the case of enforceable contracts if the optimization behaviour of both the worker and the landlord are considered. There have been several attempts, however, to establish a rationale for the existence of the share contract under certainty. This section examines the logic of these attempts.[3]

Bargaining models

Bell and Zusman (1976) consider the Nash bargaining solution of N homogeneous tenants and a single non-cultivating landlord. Tenants contemplate whether to enter into the contract with a landlord or work outside at the wage-rate of W per unit of effort, L. They consider the symmetric bargaining situation, which permits the assumption that $N - 1$ tenants have already made the tenancy contract with α and β and the remaining tenant is bargaining with the landlord. Note that, in their original model, the fixed payment does not exist but we include it here in order to show that the fixed-rent contract dominates in equilibrium.

If the tenant decides to enter into the tenancy contract, his income (Y^1) will be

$$Y^1 = \alpha F(L,H) + \beta + W(\bar{L} - L), \qquad (3.11)$$

where $\bar{L}$ is his predetermined total work-effort and the last term is his wage earned in the outside labour market.

The corresponding income of landlord (y^1) is defined as

$$y^1 = (N - 1)[(1 - \bar{\alpha}) F(\bar{L}, (\bar{H} - H)/(N - 1)) - \bar{\beta}] \\ + (1 - \alpha) F(L,H) - \beta, \qquad (3.12)$$

where $\bar{L}$ is the effort of the other $N - 1$ tenants, and $\bar{\alpha}$ and $\bar{\beta}$ are contractual terms applied to them. Because of the assumption of unenforceable work-effort, the following Marshallian relation holds:

$$\alpha F_1(L,H) = \bar{\alpha} F_1(\bar{L},(\bar{H} - H)/(N - 1)) = W. \qquad (3.13)$$

On the other hand, if the tenant does not enter into the tenancy contract, his income (Y^2) and the income of landlord (y^2) become, respectively

$$Y^2 = W\bar{L} \qquad (3.14)$$
$$y^2 = (N - 1) [(1 - \bar{\alpha}) F (\bar{\bar{L}},\bar{H}/(N - 1)) - \bar{\beta}], \qquad (3.15)$$

where the effort of other tenants $\bar{\bar{L}}$ will differ from $\bar{L}$ because of the difference in land inputs per tenant. The income differentials ($Y^1 - Y^2$) and ($y^1 - y^2$) are positive if the landlord and the tenant are rational, and represent the sources of their bargaining power. The Nash bargaining solution is obtained by maximizing the product of $(Y^1 - Y^2)$ and $(y^1 - y^2)$ with respect to α and β:

$$\max_{\{\alpha,\beta\}} M = (Y^1 - Y^2)(y^1 - y^2). \tag{3.16}$$

The optimum conditions are

$$\partial M/\partial \alpha = F[y^1 - y^2) - (Y^1 - Y^2)] + (1 - \alpha) F_1(L,H)$$
$$(Y^1 - Y^2)(\partial L/\partial \alpha) = 0 \tag{3.17}$$

$$\partial M/\partial \beta = (y^1 - y^2) - (Y^1 - Y^2) = 0, \tag{3.18}$$

where $(\partial L/\partial \alpha)$ is obtained from equation 3.13,

$$\partial L/\partial \alpha = -F_1(L,H)/\alpha F_{11}(L,H), \tag{3.19}$$

which is positive by the assumption of positive but decreasing marginal product of effort.

Focusing only on 3.17 and substituting a variety of supposedly realistic production parameters, Bell and Zusman found that α takes remarkably stable values centring on 0.5 in equilibrium. This result appears significant in view of the fact that 50:50 sharing under share contract is so common in practice. However, by substituting equation 3.18 into 3.17 it can be readily seen that $\alpha = 1$ holds in equilibrium if the option of leasehold contract is allowed.[4] This result can be anticipated, since under the assumption of production certainty the incentive problem alone causes the inefficiency and both the contracting parties can gain from wiping out such inefficiency by adopting the leasehold contract. Therefore, as in the case of the traditional Marshallian argument, the theoretical result of Bell and Zusman that the share contract exists under the unenforceable contract is derived from their restrictive assumption that excludes the option of leasehold contract.

Self-selection models

The self-selection model represents another approach to explaining the existence of share contract under certainty. In the model originally developed by Hallagan (1978), share contract is considered to exist as one of the contract arrangements from which workers themselves can choose so as to utilize in the best way their entrepreneurial ability in the absence of landlords' information on their abilities.

The Hallagan model is illustrated in Fig. 3.2 with the horizontal axis measuring the worker's input of entrepreneurial ability and the vertical axis measuring his income. The production function (F) is treated as a function of entrepreneurial ability (e) alone.

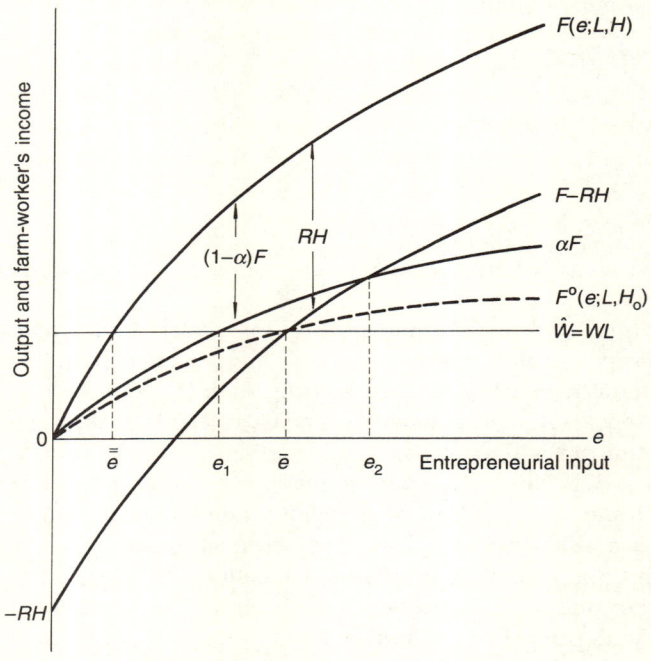

FIG. 3.2 Self-selection model of contract choice

This treatment assumes that L and H are equal among workers. It is implicitly assumed that the contract is somehow enforceable with respect to L. The three options that a worker faces are: (a) to earn a fixed-wage income ($\hat{W} = WL$) by being employed as a wage-labourer, (b) to earn a share of output (αF) under share contract, and (c) to choose the fixed-rent contract and receive the surplus of output from rent ($F - RH$). The worker whose ability is below e_1 will choose wage employment. If his ability lies between e_1 and e_2, he will choose the share contract. Further, if his ability is higher than e_2, he will choose the fixed-rent contract. Thus, Hallagan

Optimum Contract Choice 43

argues that share contract can exist as a part of the self-selection mechanism that improves the utilization of worker's entrepreneurial ability.

However, the Hallagan model has a critical shortcoming in that the worker's optimization alone is considered, neglecting the landlord's. If the landlord tries to maximize his income, he will not accept the share contract with the worker whose ability lies between e_1 and e_2 because the fixed rent (RH) is higher than the share rent ($(1 - \alpha)F$). Thus, if the landlord has no information on the ability of workers, he will propose the fixed-rent contract alone and let the workers choose either that or wage employment. Allen (1982) proved that Pareto optimality can be established in that way. In short, the Hallagan model fails to provide a rationale for the existence of share contract under certainty.[5]

A self-selection model that assumes a setting similar to Hallagan's but incorporates the landlord's optimization behaviour was developed by Allen (1985). In his model, a worker's self-selection among contractual options is used by the landlord as a device to prevent the worker's default on rent payments. In terms of Fig. 3.2, workers whose ability is lower than $\bar{e}$ would prefer wage employment to fixed-rent tenancy so long as the payments of RH are enforceable. However, if the possibility is open for workers to default on rent payments and escape to another location, those whose ability lies between $\bar{\bar{e}}$ and $\bar{e}$ may also enter the fixed-rent tenancy contract. To prevent this from happening, the landlord limits the amount of land rented out to people of unknown ability to a sufficiently small size ($H_o < H$) to ensure that low-ability workers prefer to be wage-labourers. If H_o is set to shift down the production function to $F^o(e; L, H_o)$ in Fig. 3.2, only the workers whose ability is higher than $\bar{e}$ will prefer to become tenants. They are supposed to be given more land after their ability is evident from their performance in the initial period.

A worker's income for the initial screening period is represented by

$$Y_o(e) = F(e, H_o) - RH_o. \tag{3.20}$$

Land area rented out to the high-ability worker in the subsequent period is determined so as to maximize

$$F(e, H) - RH \tag{3.21}$$

with respect to H subject to the incentive compatibility constraint of

$$RH \leq \delta\{[F(e,H) - RH] - Y_o\}, \qquad (3.22)$$

where δ is a future discount factor with respect to consumption. Equation 3.22 means that the loss in the worker's earning from leaving the present landlord and being rescreened by another landlord should not be smaller than his gain from defaulting (RH). The penalty represented by the right-hand side of equation 3.22 works as a whip to enforce the contract with respect to rent payment.

If a worker's ability is high enough to satisfy inequality in equation 3.22, the fixed-rent contract will be chosen. However, if his ability is barely sufficient to establish equality, the amount of land is rationed so as to satisfy the following relation derived from equation 3.22:

$$RH = (\delta/(1 + \delta))\, F(e,H) - (\delta/(1 + \delta))\, Y_o. \qquad (3.23)$$

Allen claims that at this level of tenant's ability, the share contract arises with $\alpha = (\delta/(1 + \delta))$ and $\beta = -(\delta/(1 + \delta))\, Y_o$. He further claims that if the discount factor is close to unity, as one may expect, the corresponding output share is around one-half.

But equation 3.23 indicates that at this level of a worker's ability, both the landlord and the worker are indifferent between the share contract and the fixed-rent tenancy with a rent of R per hectare and the rationing of land at a level smaller than that desired by the worker. This conclusion is similar to one drawn from the models discussed in Section 3.1; that is, both share and fixed-rent contracts can achieve the same resource allocation and income distribution under certainty.

Accordingly, the Allen model fails to provide a positive reason for the existence of share contract. Moreover, if the default risk is the sole problem of tenancy, the first-best solution would be achieved by requiring the payment of the fixed rent in advance. In fact, such advance payment of fixed rent is widely practised even in poor agrarian economies (Bharadwaj and Das 1975; Cohen 1983; Dow 1984; Bell and Sussagkarn 1985; Morooka and Hayami 1989). Therefore, the basis on which Allen claims the adoption of a 50 : 50 share tenancy is unsatisfactory both theoretically and empirically.[6]

Land-quality transaction-cost models

Another line of approach to find a rationale for the existence of share tenancy is to consider the cost of enforcing the terms of contracts with respect to inputs other than labour, which is in a trade-off relation with the enforcement cost of labour.

Murrel (1983) and Datta et al. (1986) and, less explicitly, Alston et al. (1984) suggested a 'transaction-cost model' with respect to the management of farm land (and capital tied to land such as fences and farm ditches). Indeed, depletion of soil fertility due to land mismanagement by tenants has long been known to be a serious problem in tenancy contracts (Wallace and Beneke 1956). It is reasonable to assume that the incentive for the tenant to gain from abuses of farm land and capital becomes larger as σ becomes larger. Therefore, the cost of enforcing the terms of the contract with respect to work-effort, or the efficiency loss arising from the Marshallian misallocation of work-effort, is inversely related to the cost with respect to land-quality management, as illustrated in Fig. 3.3. If, in fact, both the cost curves of monitoring work-effort and land quality are convex and independent as shown in Fig. 3.3, α^* will be determined at the point where the slopes of the two cost curves are equal in absolute terms so that the total transaction cost for the landlord is minimized.[7] An equilibrium can be reached if β is adjusted properly for a given α^* so as to assure the tenant's income at the level of his reservation income or utility. This equilibrium is not Pareto-optimal, and its solution is expected to have the same properties as those of the unenforceable contract under certainty shown in Table 3.1.

This land-quality transaction model offers a consistent explanation for the existence of share tenancy under the assumption of unenforceable contract and the absence of risk-aversion under uncertainty. However, it applies only to the situation in which the tenancy contract is made for a short period during which the effect of land mismanagement is not so evident. If it is planned to renew the contract, it is to the tenant's advantage not to shirk in order to establish a reputation to maximize his future income, as will be discussed in Section 4.3.

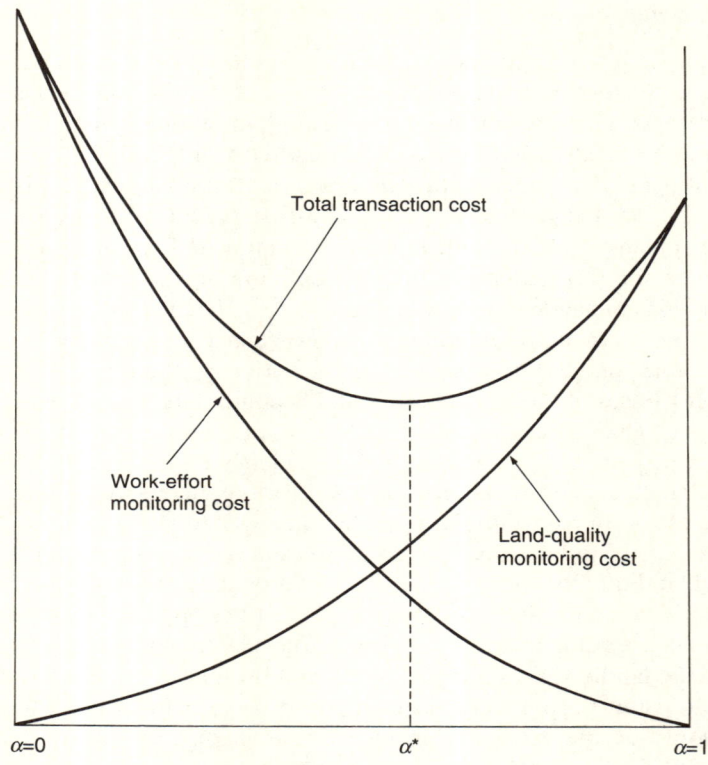

FIG. 3.3 The land-quality transaction-cost model of contract choice

3.3 Optimum Contract Choice under Uncertainty

This section attempts to show that, in the presence of risk-aversion under conditions of uncertainty, the existence of share contract can be explained in regard to its role in risk-sharing. In the analysis that follows, the basic model of contracts developed in Chapter 2 is considered, wherein the problem of enforcement applies only to the worker's effort.[8]

Enforceable contract

In the case of the landlord's optimization problem specified in equation 2.13, where the worker's effort is enforceable, the maximization of the landlord's utility, subject to the constraint that the worker is assured of receiving his reservation utility, establishes an equality of the expected marginal land productivities between owner-farmed and worker-cultivated land:

$$f_2 - F_2 = 0, \quad (3.24)$$

which is the condition of efficient resource allocation. Equation 3.24 implies that the effort/land ratios are the same for owner-farmed and worker-cultivated land under the assumption of the linear homogeneous production function. Such efficient resource allocation is realized because if the effort/land ratios are different the landlord can increase the expected output and, hence, his income by simply reallocating L, H, ℓ, and h under the enforceable contract. This relation holds irrespective of which combination of α and β is chosen. Therefore, in the case of enforceable contracts, both inputs and expected outputs per hectare will be the same for different types of tenancy and labour contract.

Having established efficient resource allocation, the landlord can manipulate α and β to achieve optimum risk-sharing. In equilibrium, the following condition of optimal risk-sharing is obtained:

$$Eu_1\theta/Eu_1 - EU_1\theta/EU_1 = 0, \quad (3.25)$$

which implies that the marginal rates of substitution between risky income and riskless income are the same for the worker and the landlord. If the contracting parties are risk-averse (i.e. $u_{11} < 0$ and $U_{11} < 0$), then u_1 and U_1 are negatively correlated with θ, which implies that both ratios in equation 3.25 are less than unity (Stiglitz 1974).[9] By approximating the utility functions in equation 3.25 with respect to θ by the Taylor expansion up to second-order terms and utilizing equation 3.24, the following relation is obtained:

$$\alpha = \frac{aH}{a(H - h) + AH}, \quad (3.26)$$

where $a = -(u_{11}/u_1)$ and $A = -(U_{11}/U_1)$ are the measures of absolute risk-aversion of landlord and worker respectively, evaluated at the mean income. If the landlord is risk-neutral ($a = 0$) and the worker is risk-averse ($A > 0$), the fixed-wage labour contract ($\alpha = 0$) by which all the risk is shouldered by the landlord will be chosen (Stiglitz 1974; Mitra 1983). If the landlord is risk-averse ($a > 1$) and the worker is risk-neutral ($A = 0$), the fixed-rent contract ($\alpha = 1$) will be chosen in which all the risk is shouldered by the worker, so long as the landlord does not farm his land himself ($h = 0$) (Otsuka and Murakami 1987). If both the landlord and the worker are risk-averse ($a > 0$, $A > 0$), the share contract ($0 < \alpha < 1$) will be chosen for the sake of risk-sharing (Stiglitz 1974; Sutinen 1975; Hirshleifer and Riley 1979). Thus, the positive reason for the existence of share tenancy is identified as risk-sharing under the condition of uncertainty.

An equilibrium condition of income distribution can also be derived as follows:

$$\alpha F + \beta = LF_1 + \beta[1 - (Eu_1/Eu_1\theta)]. \qquad (3.27)$$

The left-hand side of the equation represents the expected income of the worker; the first term on the right-hand side represents the expected contribution of the worker's effort to output; and the second term is positive so long as β is negative (fixed-rent payment) and the landlord is risk-averse. This second term may be considered a risk premium that the landlord pays to the worker in addition to the worker's contribution to output for the landlord's receipt of non-risky fixed rent ($-\beta$). This risk premium becomes larger as α and ($-\beta$) become larger. Because of this risk premium, the fixed-rent tenant's income is expected to be higher than the share tenant's income. Correspondingly the landlord's rent income, which is the difference between output and the worker's income, is smaller under a fixed-rent than under a share contract.

The implications for resource allocation and income distribution of a share contract compared with other forms of land tenure under enforceable contract with uncertainty are summarized in the third row of Table 3.1.

Can a combination of contracts be a substitute?

Identification of the share contract as a means of achieving optimum risk-sharing has been questioned on the ground that the same distribution of risk can be achieved by a linear combination of fixed-wage and fixed-rent contracts under the assumptions of constant returns to scale and no transaction cost (Stiglitz 1974; Newbery 1975a, 1977; Reid 1976b; Newbery and Stiglitz 1979). This is because a worker incurs no risk under the fixed-wage contract and he assumes the whole risk under the fixed-rent contract, so that the appropriate mix of the two can result in a contract equivalent to any share contract.

Stiglitz and Newbery argued that a positive reason for the existence of share tenancy should be sought in scale economies in the workers' production function arising from indivisibilities of inputs such as their entrepreneurial ability, family labour, and draught animals for which a market is absent or inactive. However, as Allen (1984) argued, even if scale economies exist, the mixture of wage and rent contracts can be equivalent to a share contract when enforcement cost is zero, which ensures the same resource allocation between the mixed and the share contracts.

This equivalence theorem was originally formulated to disprove Cheung's (1969) argument that share tenancy will be chosen in preference to fixed-rent and fixed-wage contracts for its risk-sharing advantage, even if the enforcement cost is highest under share contract. As the equivalence theorem implies, Cheung's argument is incorrect, since the combination of the fixed-rent and the fixed-wage contract dominates the share contract in such cases.

There is little doubt that in the real world contract choice will be made with consideration of both risk and transaction costs. For the same risk-sharing, share contracts will be preferred to a mixture of contracts if the transaction costs involved in the former are smaller than in the latter. This seems to be often the case since the fixed-wage contract provides no work-incentive for the worker so that the very high cost of its enforcement is likely to outweigh the low enforcement cost of the fixed-rent contract. The empirical relevance of the equivalence theorem, therefore, is questioned.

Unenforceable contract

If enforcement of work-effort involves significant cost under uncertainty, an increase in α enhances the work-incentives only at the cost of increasing the portion of risk shouldered by the risk-averse worker. In such a situation, share contracts ($0 < \alpha < 1$) will be chosen so long as the workers are risk-averse, even if work-incentives are sacrificed (Stiglitz 1974; Hiebert 1978). However, even though the worker's incentives increase with α, the equilibrium effort per unit of land can be lower or higher under a share contract than under cultivation by the owner. This is because, when the landlord increases α, he will decrease H so as to maintain the equality of the worker's utility with his reservation utility. Whether the incentive effect of an increase in α on L outweighs the disincentive effect of a decrease in H in the worker's effort reaction function (equation 2.11) depends on the elasticity of substitution between L and H (σ). In fact, the difference in the expected marginal land productivities between farming by the owner and cultivation by the worker depends on σ in the following manner:

$$f_2 - F_2 = \alpha(1 - \alpha)(F^2 F_2/F) \, EU_1 \theta((\sigma - 1)/\sigma) \\ (\partial^2 EU/\partial L^2). \qquad (3.28)$$

Inefficiency in resource allocation emerges generally because an equilibrium condition is not zero and, hence, the equality in the expected marginal productivities of land does not hold between worker's cultivation and farming by the owner except in the special case in which the elasticity of substitution is unity.[10] In general, $f_2 > F_2$ for $\sigma < 1$ or $f_2 < F_2$ for $\sigma > 1$, implying that work-effort and therefore output per hectare of land under share contract are lower than those of owner-farming and fixed-rent tenancy for $\sigma < 1$ but are higher for $\sigma > 1$.

In this situation risk-sharing will not be optimal because

$$Eu_1 \theta / Eu_1 - EU_1 \theta / EU_1 = -(1 - \alpha) F_1 \, (Eu_1 \theta / Eu_1) \\ (EU_1 \theta / EU_1)(\partial L / \partial \beta) \\ + Eu_1 \theta / Eu_1 [(F_2 - F_2)/\alpha F_2] \quad (3.29)$$

is not zero unless workers are risk-neutral, and, hence, fixed-rent tenancy ($\alpha = 1$) is chosen. Unlike the case of enforceable contract, optimum risk-sharing cannot be achieved under the share contract because of the trade-off between providing work-incentives and attaining the desired risk-sharing.

An equilibrium condition of income distribution in the unenforceable contract case is given by

$$\alpha F + \beta = LF_1 + \beta \, [1 - (Eu_1/Eu_1 \theta)] + H(F_2 - f_2), \quad (3.30)$$

which is different from equation 3.27 because of the third term on the right-hand side. This term is known from equation 3.28 to be non-positive for $\sigma \leq 1$. In this case, the expected income of the worker ($\alpha F + \beta$) tends to be smaller than the expected contribution of his effort to output (LF_1) because of the implicit penalty on lower marginal product on tenanted land. The same logic as applied to equation 3.27 establishes a relation in which the worker's income is larger and land rent is smaller under the fixed-rent than under the share contract in so far as $\sigma \leq 1$. However, if $\sigma > 1$, relative magnitudes of the worker's income and land rent between the share and the fixed-rent contracts become indeterminate.

The implications for resource allocation and income distribution of the share contract compared with other forms of land tenure under unenforceable contract with uncertainty are summarized in the fourth, fifth, and sixth rows in Table 3.1.

3.4 On the Relevance of Alternative Assumptions

As discussed in Chapter 1, monitoring of the worker's effort by the landlord in a spatially dispersed and ecologically diverse environment is not an easy task, while insuring against a subsistence crisis is likely to be a crucial consideration for poor landless people in agrarian economies. Thus, the assumption of the unenforceable contract in the presence of risk-aversion seems more realistic. Moreover, this model justifies the existence of share contract in the theoretically most consistent manner among those we have reviewed in this chapter. Yet these considerations do not immediately deny the value of the enforceable contract model. As Friedman (1953) puts it, the usefulness of simplifying assumptions should not be judged solely on the reality of the assumptions *per se* but ultimately on the empirical predictability of their implications. Indeed, the recent trend in the agency literature is to seek richer contractual devices which can improve contractual efficiency under conditions of unenforceable contract, particularly within a framework of long-term contracts where reputation plays a significant role in enforcement. Thus, there is a possibility that the one-period model of enforceable contract does provide the shorthand solution which may sufficiently well approximate the equilibrium made possible by the use of a complex set of contractual devices, even if this assumption is unrealistic.

The truth, however, will lie between the two polar cases of totally unenforceable and enforceable contracts. The more relevant assumption, therefore, is that of imperfectly enforceable contracts, which includes those two extreme assumptions as special cases. In the next two chapters we will review the theories of long-term and interlinked contracts focusing on a possible enforcement mechanism latent in less developed agrarian economies. The aim is to bridge the gap between the simplified picture of land and labour contracts of the basic one-period model and the more complicated, enduring contractual relations described by casual observations. In particular, we will attempt to build a general model of long-term agrarian contracts incorporating a monitoring function of the worker's effort by the landlord as well as additional incentive mechanisms based on the role of reputation.

Appendix: Derivation of the Optimum Conditions

How the optimum conditions of the basic one-period model, which was formulated in Chapter 2 and examined in Chapter 3, are derived is shown here.

Unenforceable contract

The problem of optimization by the landlord can be solved by applying the Lagrangean method (Otsuka et al. 1993). In the analysis of share tenancy, however, it has been customary first to specify the relations among strategic variables from the constraints in order to facilitate interpretation of the nature of equilibrium. Here, H and L are specified as functions of α and β based on equations 2.8 and 2.10. Differentiation of equation 2.8 with the use of enveloped theorem results in

$$(\partial H/\partial \alpha)_V = -F/\alpha F_2 < 0, \tag{A1}$$
$$(\partial H/\partial \beta)_V = -EU_1/EU_1\theta\alpha F_2 < 0. \tag{A2}$$

The above equations show that, when the terms of contract improve for the worker with increased values of α and β, H must decrease if EU is maintained at a level equal to V.

Total differentiation of equation 2.10 leads to

$$\begin{aligned}(\partial L/\partial \alpha) &= -1/EU_{LL}\{[EU_1\theta F_1 + EU_{11}\theta^2\alpha F_1 F + EU_{12}\theta F] \\ &\quad + [EU_1\theta\alpha F_{12} + EU_{11}\theta^2\alpha^2 F_1 F_2 + EU_{12}\theta\alpha F_2] \\ &\quad (\partial H/\partial \alpha)_V\} \\ &= -EU_1\theta F_1/EU_{LL}[(\sigma - 1)/\sigma)],\end{aligned} \tag{A3}$$

$$\begin{aligned}(\partial L/\partial \beta) &= -1/EU_{LL}\{[EU_{11}\theta\alpha F_1 + EU_{12}] \\ &\quad + [EU_1\theta\alpha F_{12} + EU_{11}\theta^2\alpha^2 F_1 F_2 + EU_{12}\theta\alpha F_2] \\ &\quad (\partial H/\partial \beta)_V\},\end{aligned} \tag{A4}$$

where $EU_{LL} = \partial^2 EU/\partial L^2 < 0$, the negative sign being due to the second-order condition of the worker's utility maximization, and σ is the elasticity of substitution between L and H. Work-effort (L) decreases with an increase in α if $\sigma < 1$, and vice versa if $\sigma > 1$ (Newbery and Stiglitz 1979; Braverman and Srinivasan 1981).

Under the functional constraints represented by equations A1–A4, the landlord wishes to maximize

Optimum Contract Choice

$$\max_{\{\alpha, \beta, N, \ell\}} Eu(y, \ell) \equiv u^*, \quad (A5)$$

which yields the following first-order conditions:

$$u_\alpha^* = -NEu_\ell\theta\{F - (1-\alpha)F_1(\partial L/\partial\alpha) - [(1-\alpha)F_2 - f_2](\partial H/\partial\alpha)_v\} = 0, \quad (A6)$$

$$u_\beta^* = -NEu_1 + NEu_1\theta\{(1-\alpha)F_1(\partial L/\partial\beta) + [(1-\alpha)F_2 - f_2](\partial H\partial/\partial\beta)_v\} = 0, \quad (A7)$$

$$u_N^* = Eu_1\theta[(1-\alpha)F - Hf_2] - Eu_1\beta = 0, \quad (A8)$$

$$u_\ell^* = Eu_1\theta f_1 + Eu_2 = 0. \quad (A9)$$

Equation 3.28, which expresses the difference in the expected marginal product of land between owner-cultivated and tenanted land, is obtained by inserting equations A1 and A3 into equation A6. Equation 3.29, which specifies the second-best risk-sharing rule, is derived from equations A2, A4, and A7. Finally, equation 3.30 concerning the condition of the income distribution is obtained from equation A8 with the application of Euler's theorem.

It can be easily ascertained that equation 3.5, analysed in the special case of certainty in Section 3.1, is derived from equations A6 and A7.

Enforceable contract

In the enforceable contract case, H can be solved from equation 2.8 as a function of α, β, and L. While the partial derivatives of H with respect to α and β are formally identical to equations A1 and A2, the partial effect of L can be expressed as

$$(\partial H/\partial L)_v = -(EU_1 Q\alpha F_1 + EU_2)/EU_1\theta\alpha F_2. \quad (A10)$$

Under the functional constraints represented by equations A1, A2, and A10, the landlord wishes to maximize

$$\max_{\{\alpha, \beta, L, N, \ell\}} Eu(y, 1) \equiv \bar{u}, \quad (A11)$$

which yields the following first-order conditions:

$$\bar{u}_\alpha = -NEu_1\theta\{F - [(1-\alpha)F_2 - f_2](\partial H/\partial\alpha)_v\} = 0, \quad (A12)$$

$$\bar{u}_\beta = -NEu_1\{1 - \theta[(1-\alpha)F_2 - f_2](\partial H/\partial\beta)_v\} = 0, \quad (A13)$$

$$\bar{u}_L = NEu_1\{(1-\alpha)F_1 + [(1-\alpha)F_2 - f_2](\partial H/\partial L)_v\} = 0, \quad (A14)$$

$$\bar{u}_N = Eu_1[(1-\alpha)F - Hf_2] - Eu_1\beta = 0, \quad (A15)$$

$$\bar{u}_\ell = Eu_1\theta f_1 + Eu_2 = 0. \quad (A16)$$

The optimum condition for the resource allocation expressed in equation 3.24 is obtained by substituting equation A1 in A12, whereas the optimum condition for the risk-sharing specified in equation 3.25 can be derived from equations A2, A13, and 3.24. The condition of the income distribution shown in equation 3.27 is obtained from equation A15 by applying the Euler's theorem and using equation 3.24.

For the certainty case it can be easily demonstrated that equations A12 and A13 both reduce to the same relation, $f_2 = F_2$. This implies that a unique solution does not exist for α and β under certainty, as emphasized in Section 3.1.

Notes to Chapter 3

1. Moreover, equation 3.6 can be considered as an equilibrium condition compatible with the optimizing behaviour of the worker with respect to H. See Ch. 2, n. 8.
2. Assuming that the landlord can impose the minimum output constraints under certainty, Taslim (1989) also shows that the share tenant, in fear of eviction, will work as hard as the owner-cultivator. The assumption of certainty, however, presupposes the observability of the tenant's work-effort. Under such conditions, it is trivial to obtain the efficient outcome.
3. The transaction-cost model of management-input formulated by Eswaran and Kotwal (1985*b*) offers a theoretically consistent explanation for the existence of share contracts within the framework of the one-period model under the assumption of unenforceable contract and the absence of risk-aversion under uncertainty. We evaluate critically the relevance of their model in the context of the long-term contract in Ch. 4.
4. Bell (1988) extends the Bell and Zusman model to include the risk of default by the tenant in payment of rent arising from low realization of θ. More recently Bell and Zusman (1989) developed the bargaining model under the assumptions of heterogeneous actors and interdependent contracts. In these models, while the equilibrium contract may not be leasehold, the puzzle of 50 : 50 sharing cannot be explained.
5. Newbery and Stiglitz (1979) explain the existence of share tenancy within the framework of the self-selection model consistent with the rational behaviour of landlords but with the assumption of uncertainty and risk-aversion of workers. Braverman and Guasch (1984) construct the self-selection model under 'certainty', in which β is assumed to be zero and credit plays the role of fixed payment. Differences in the

Optimum Contract Choice

credit terms induce the self-selection of contracts by workers with different abilities, but the interlinked credit contract *per se* does not justify the self-selection mechanism, since β can also be changed.

6. Moreover, there is general presumption in the tenancy literature that the quality of tenants is known to landlords within a village economy (Bardhan 1984: 97; Bell 1988).
7. As Alston and Higgs (1982) pointed out, however, there may be economies of scope associated with supervision so that the joint costs of monitoring tenant's labour input and land abuse may be less than the sum of those two curves as shown in Fig. 3.3.
8. A model parallel to the tenancy models reviewed here is developed for the analysis of the recollectivization of agriculture by Carter (1987).
9. Covariance between u_1 and θ is expressed as $E(u_1 - Eu_1)(\theta - 1) = Eu_1\theta - Eu_1$. Since this is negative if the landlord is risk-averse, $(Eu_1\theta/Eu_1) < 1$ holds. The similar relation holds for the worker.
10. If there are more than two inputs, the difference in the expected marginal productivity of inputs depends on the direct elasticities of substitution among inputs in a complicated way. The difference is zero, however, when all the direct elasticities of substitution are unity (i.e. the case of the Cobb–Douglas production function). See Otsuka and Murakami (1987) for the case of three inputs.

4

Long-Term Contracts

IN the basic one-period contract model, a contractual agreement is supposed to be made at the beginning of a crop season and payment is made at the end. A difficulty might arise when one or both of the contracting parties find it in their interests to change the rules of the contract *ex post*. An advance payment does not necessarily solve the problem, since an incentive is then created for those who have received it in advance to claim that the payment was not made. The problem of such strategic default and *ex post* bargaining, however, can be solved by signing a contract which explicitly states the payment obligations and recording the actual payments. The enforcement of work-effort, if it is specified in the contract, is more problematic, since the objective evidence of shirking is hard to come by. In short, the source of the problem lies in the difficulty of verifying breach of contract to a third party.

An implicit assumption of the enforceable contract of our basic model is that those who may deviate from the agreement can be penalized sufficiently to deter them from such a course. However, the way in which the penalty is imposed may affect the behaviour of contracting parties. Indeed, Mirrlees (1974) has shown that it is possible for the principal to elicit approximately the first-best level of effort from the agent by threatening him with an arbitrarily severe penalty whenever a very small output, assumed to occur only when the agent shirks, is observed. Thus, if the agreement on minimum output can be enforced, the explicit inclusion in the contract of a clause on the severity of penalties may ensure the first-best outcome.

The assumption that a severe penalty can be imposed, however, is not realistic. In fact, the worker may desert to other landlords or to other jobs to avoid such a penalty. In general, the existence of alternative job or contract opportunities tends to limit the amount of a penalty, unless a third party such as a court exists to enforce the contract and to execute the punishment. Even if a court exists as a last resort, litigation is often prohibitively costly. In such a case, a self-enforcing mechanism must be sought.

One such mechanism is the long-term contract, in which a party not honouring a contract will suffer a loss of reputation and this will reduce his value as someone with whom to contract in the future. Such a consideration has attracted a great deal of attention in recent agency models and we first review briefly the logic of their arguments. Then in subsequent sections we examine the role of reputation in contract enforcement in long-term labour employment and land-tenancy models. We attempt to show how the increased options of penalties and rewards in long-term contracts contribute to the enhancement of work-incentives, thereby increasing the efficiency of the unenforceable contract.

4.1 Theories of Long-Term Contracts

Reputation plays a central role in recent microeconomic theories, particularly those in a game-theoretic framework (Wilson 1985). The key idea is that the evolution of a person's reputation depends on his past actions and that a reputation, thus formed, affects his future economic opportunities. In other words, each agent needs to evaluate the effects of his current actions on his reputational capital.

Radner (1981) and Rubinstein and Yaari (1983) consider the reputation of an agent through past performance within the context of a long-term agent–principal model, in which the same single-period contract is repeated over an infinite, or approximately infinite, number of periods. Since the effect on production of random factors is filtered out in the very long run, a penalty can be imposed if the average output in the past differs from the first-best expected output. For fear of future penalty, the agent is forced to apply the first-best effort-level in each period.[1] Moreover, this contract is self-enforcing so long as the output is verifiable. Even if the contract period is finite, it is in the interest of the principal to relate current-period reward not only to current-period output but also to outputs in the past in order to mitigate the randomness associated with single-period output (Lambert 1983). In this situation, the myopic shirking behaviour of the agent tends to be deterred, though not eliminated, since he must take into account the effect of his action not only on his current earnings but also on subsequent earnings.

In the theory of the implicit labour contract, more explicit treatment of the effect of reputation is considered within the context of a finite-period model (Holmstrom 1981, 1983; Carmichael 1984). For an employer reputation is important because it affects his future costs when hiring workers. A similar concern applies to the workers because their reputation affects the probability of lay-offs and the terms of future contracts. If failure to take the actions prescribed by the contract results in a loss in reputation that causes such a reduction in welfare as to deter offenders, the contract becomes effectively enforceable.

However, as Bull (1987) argues, for reputation to be an effective enforcer, information about contract-breaching behaviour must be efficiently transmitted to outsiders (potential employees and employers) as well as insiders (current employees). If the information flow in the labour market is imperfect, the incentives to breach the contract will remain. By adopting the more reasonable assumption that the information flow is perfect inside the firm, Bull proposes the overlapping-generation model of the labour contract, in which an employer's dishonesty and maltreatment of older-generation employees are punished by reduced profits due to a decline in the work morale of younger-generation employees.

In agrarian economies a farm population, except for seasonal migrant labourers, is usually long settled in a village community and people know one another quite well through an efficient mouth-to-mouth communication network (Section 1.2). In such circumstances, reputation certainly matters because contract violators will soon become known in the community, thereby affecting the possibility of continuing or entering a contract and, in extreme cases, even triggering ostracism. Thus, there is good reason to believe that a built-in self-enforcing mechanism exists in tenancy and permanent-labour contracts in less developed agrarian economies.

In the next section we attempt to build a model of long-term land-tenancy and labour contracts which incorporates a monitoring function of the farm-worker's effort and the role of reputation in contract enforcement. We demonstrate that the larger the expected utility loss arising from loss of reputation in the community, the smaller the efficiency loss of agrarian contracts. Moreover, we show that the cases considered by existing models of long-term (permanent) labour contract as well as one-period tenancy models (reviewed in Section 3.3) correspond to the special cases in our model.

4.2 Theories of Long-Term Agrarian Contracts[2]

Eswaran and Kotwal (1985a) have recently proposed a model of the long-term permanent-labour contract in which shirking by a permanent labourer results in the loss of his reputation and lucrative future employment opportunities. Focusing exclusively on the choice between permanent- and casual-labour contracts, Eswaran and Kotwal (E-K) attempt to show the existence of the fixed-wage permanent-labour contract in equilibrium, which is designed to elicit loyal effort from permanent labourers by assuring them of utilities higher than their reservation utilities. E-K assume a priori that only the fixed-wage contract is offered to farm-workers. Binswanger and Rosenzweig (1986) and Rosenzweig (1988c) conjecture, however, that such a contract will never be observed if the opportunity exists to lease out the land because of the superior work-incentives provided by the tenancy contracts. This conjecture would be correct if the choice of contract were considered endogenous.

While E-K assume that the crop season consists of two periods, we consider only one period since the two-period assumption is not essential for their main theoretical result.[3] We maintain, however, their assumption that there are two categories of labour inputs: casual labour engaging in such simple, easily monitorable tasks as weeding and harvesting, and permanent labour engaging in such care- and knowledge-intensive tasks as land preparation and fertilizer application.

For the sake of simplicity, it is assumed that a landlord owns a certain area of land and employs only one permanent worker together with N_c number of casual labourers. The production function is defined as

$$Q = \theta F(L_p, N_c L_c), \qquad (4.1)$$

where L_p and L_c refer to the effort of permanent and casual labour inputs respectively, and land input is suppressed to simplify the exposition. For brevity, we use the same notation F to denote the production function, which is subject to constant returns in L_p, $N_c L_c$, and the fixed factor, H. All labourers are assumed to have identical abilities and preferences. The thrust of the E-K model is that even in such a situation the two-tier labour market consisting of distinct labour classes (i.e. permanent and casual labourers) emerges.

The reward to the permanent worker (Y_p) consists of a share (α) of net output and the fixed payment (β):

$$Y_p = \alpha(Q - N_c W_c L_c) + \beta, \tag{4.2}$$

where W_c denotes wage-rate per unit of casual labourer's effort, which is assumed to be determined in the competitive market. The income of a casual labourer (Y_c) is defined as

$$Y_c = W_c L_c. \tag{4.3}$$

Note that wage payments to casual labourers, which are in proportion to L_c, are feasible because L_c is assumed perfectly and immediately monitorable. Since the permanent labourer may be more legitimately described as a tenant when $\alpha > 0$, we will call him by the term 'long-term farm-worker', which subsumes both the permanent labourer and the tenant.[4]

As in the E-K model, the landless farm-worker is required to promise effort at least equal to a certain minimum level, $\bar{L}$. If the violation of condition $L_p \geq \bar{L}$ is discovered by the landlord, the renewal of the contract is refused. Further, the worker is assumed to lose his reputation as a diligent worker in his community and is forced to drop to the rank of casual labourer in subsequent periods.[5] E-K assume that by nature of the tasks L_p is unmonitorable until harvest, implying that the actual level of L_p can be identified only after Q is observed. The assumption of *ex post* perfect monitorability, however, seems too restrictive in the presence of risk factor θ. We introduce a monitoring function of the worker's effort which specifies the probability of detecting his shirking. That is, we assume that the worker's shirking, if committed, will be detected only with probability of ψ.[6] We also assume that ψ depends positively on the degree of shirking ($\bar{L} - L_p$) and supervision time expended by the landlord(s):

$$\psi = \psi(\bar{L} - L_p, s), \quad \text{for } (\bar{L} - L_p) > 0, \tag{4.4}$$

which is assumed to be concave and characterized by non-negative partial derivatives with respect to ($\bar{L} - L_p$) and s, i.e. $\psi_1, \psi_2 \geq 0$. We assume that if it is prohibitively costly for the landlord to monitor the worker's effort, $\psi = \psi_1 = \psi_2 = 0$ always hold. Otherwise, $\psi_1, \psi_2 > 0$ is obtained.

Regardless of whether the long-term worker observes the stipulated minimum level of effort, however, the landlord is supposed to be committed contractually to pay him Y_p. Without such a commitment, the landlord cannot be relied on to pay this sum even to honest workers who contribute more than $\bar{L}$. The implicit assumption here is that the fear of loss of reputation and the consequent difficulty of finding conscientious long-term workers obliges the landlord to live up to his commitment.

It is assumed that the same contract is repeated infinitely, unless the worker's shirking is detected by the landlord.[7] Then the long-term worker's optimization problem is formulated as

$$J = \max_{\{L_p\}} EU\ (\alpha Q^* + \beta, L_p) + \delta\ [\psi J' + (1 - \psi_1)J], \qquad (4.5)$$

where the asterisk denotes the optimum value for the worker given, α, β, $\bar{L}$, and s; δ is the utility discount factor; and J' corresponds to the worker's lifetime reservation utility to be realized when he becomes a casual labourer. The second term in equation 4.5 shows his discounted expected future utility. We assume that there are many homogeneous landless farm-workers in the village economy so that J' is exogeneously given to each worker.

The first-order condition for the worker's optimization problem is

$$EU_1 \alpha \theta F_1 + EU_2 + \delta(J - J')\psi_1 \leq 0, \qquad (4.6)$$

where the first term shows the worker's utility gain from higher output; the second term represents the utility loss associated with greater effort; and the last-term shows the change in expected penalty on shirking arising from the loss of his current job when the worker changes L_p, provided that $(J - J') \geq 0$. Unlike the one-period model, the fixed-wage worker's optimum choice of effort is not necessarily zero but could be $\bar{L}$ if continuation of the contract is more beneficial than breaking it. Herein lies the advantage of the long-term over the single-period contract in that possible loss in future utilities may deter shirking by workers even under the fixed-wage contract. If such a condition is not satisfied, however, the long-term worker will breach the contract and choose to do less than $\bar{L}$.

The one-period unenforceable contract model reviewed in Chapter 3 corresponds to the case of $\psi_1 = 0$ because of the extreme difficulty of monitoring. On the other hand, the one-period enforceable contract model corresponds to the case of $\psi = 1$, in which $\bar{L}$ is set at the first-best efficient level. As a result, the worker is forced to apply effort up to the level where the sum of the first two terms in equation 4.6 becomes negative. These two cases are somewhat unrealistic, and the case of imperfect monitoring is obviously more relevant.

The landlord maximizes his expected utility with respect to α, β, N_c, s, and $\bar{L}$ subject to the reservation-utility constraint $(J \geq J')$ and to the worker's effort reaction function.[8]

The permanent-labour contract model of E-K assumes a priori that $\alpha = 0$. Then the first term in equation 4.6 becomes zero. Thus, without some form of penalty for shirking, the worker has no incentive to work. E-K characterize the permanent-labour contract equilibrium by $J - J' > 0$, on the assumption of *ex post* monitorability of the worker's effort, i.e. $\psi = 1$ for $(\bar{L} - L_p) \geq 0$. This result is easily confirmed from equation 4.6.[9] In other words, if the reward scheme does not provide adequate work-incentives, the landlord must guarantee the worker utility higher than his reservation utility in order to provide them.

It is also obvious from equation 4.6 that it makes sense for the landlord to set $\bar{L}$ and supervise the worker's effort only if $J - J' > 0$ holds, as in the labour-supervision models of Calvo and Wellisz (1978, 1979). Otherwise, the worker would not care about the supervision, since he is in a position to shift to other jobs at any time.

Yet so long as $J > J'$, the contract is second-best. As Chuma *et al.* (1990) have demonstrated, if the landlord can choose α and β to his best advantage in the absence of worker's risk-aversion, he may choose the fixed-rent contract with $\alpha = 1, \beta < 0$, and $\bar{L} = 0$, since by doing so he can attain the maximum, first-best utility without inducing the long-term worker to shirk. Therefore, although the long-term contract can provide the added work-incentives in the form of $J > J'$ the fixed-wage contract is dominated by the fixed-rent contract, as in the basic one-period model under 'certainty'. In the more general case where the risk-aversion of the worker is present and monitoring is imperfect, the share contract is chosen (Otsuka *et al.* 1993). Moreover, as in the model of E-K, $J > J'$ may hold in the share-tenancy equilibrium in order to strengthen the work-incentives of the share tenant. The utility differential, however, may be smaller, the more efficient is the supervision by the landlord, implying that the efficiency of the share-tenancy contract depends on the ability of the landlord to detect the tenant's shirking.

Theories of permanent-labour contract formulated by Bardhan (1979b, 1983) and Bell and Srinivasan (1985b) are also concerned with the possible advantage of the fixed-wage permanent-labour contract over the spot-market, casual-labour contract; there is no option of a tenancy contract. Anomalously, they assume that the work-effort of the permanent labourer is perfectly enforceable without any incentive scheme, despite the fact that in their earlier studies of share tenancy they have employed the Marshallian assumption that the tenant's work-effort is totally unenforceable (Bardhan and Srinivasan 1971; Bell and Zusman 1976; Bardhan 1977, 1979a). Assuming that a crop season consists of two periods, Bardhan (1979b) identifies the saving of recruitment cost as the advantage of the longer-term fixed-wage contract over the spot-market wage contract. This would not explain, however, why the tenancy contract, whose length is also seasonal, is not chosen. Bardhan (1983) and Bell and Srinivasan (1985b) attribute the advantage of the fixed-wage contract to the benefit of risk-sharing between the risk-averse worker and the risk-neutral landlord in the face of uncertain spot-market wages.[10] This conclusion is essentially the same as the result obtained in the basic model in Chapter 3 that if the contract is enforceable between risk-averse worker and risk-neutral landlord the fixed-wage contract is chosen. The assumption of perfect enforceability of the fixed-wage contract by Bardhan, and by Bell and Srinivasan, however, is untenable unless effective enforcement mechanism is specified.

In sum, the following testable results can be derived from our theoretical analysis in this section. First, the fixed-wage permanent-labour contract will not usually be chosen, unless tenancy is socially or legally prohibited. Secondly, where the option of tenancy is precluded, the landless farm-worker will be offered higher utility when he engages in the permanent-labour contract than when he engages in the casual-labour contract. Thirdly, resource allocation under the permanent-labour contract will be less efficient than under tenancy contracts as well as under owner-farming because permanent labour is an imperfect substitute for the family labour of tenant- and owner-farmers. Lastly, share tenancy may not be significantly inefficient if the probability of detecting shirking and the cost of losing reputation are sufficiently high. We examine these hypotheses empirically in Chapters 6 to 9.

4.3 Reputation and Contract Enforcement

When work-effort cannot be effectively monitored and enforced, the fixed-rent contract dominates the share and fixed-wage contracts in the absence of risk-aversion in our basic model as well as in the permanent-labour contract model of E-K, simply because an increase in α always improves the efficiency of the contract. In contrast, if the contracting parties are concerned with income risk and the work-effort is unenforceable, a trade-off arises between providing incentives and sharing risk that results in the choice of a share contract. This implies that if some form of the trade-off relation emerges with an increase in α, the share contract will be chosen in equilibrium. This idea has been incorporated into several recent studies of tenancy contract within the context of the one-period model. In this section, we examine the logic of these models in the light of the robustness of their assumptions on contract enforceability with respect to the lengthening of the contract period.

First, let us examine the share-tenancy model of Eswaran and Kotwal (1985b) that attempts to explain the existence of share tenancy in the absence of risk-aversion.[11] Drawing on the theory of the agricultural ladder (Spillman 1919), E-K's share-tenancy model considers a situation in which young tenants who have little experience with farm management enter into the share contract in order to secure advice from experienced landlords. In this model, the landlord is supposed to provide his management time, m, per farm-worker in addition to the fixed area of land, H. Thus the production function in the worker's cultivated area is defined as

$$Q = F(L, m), \qquad (4.7)$$

where F is a constant-returns production function, with respect to L, m, and the fixed factor H. As in the Marshallian model, the worker's optimization behaviour results in[12]

$$\alpha_1 F_1 U_1 + U_2 \leq 0. \qquad (4.8)$$

The landlord's utility function is now defined as $u(y, m)$ and his optimization behaviour with respect to m, which the worker cannot enforce, leads to

$$(1 - \alpha)F_2 u_1 + u_2 \leq 0. \tag{4.9}$$

In this model, the fixed-rent contract is unlikely to be chosen, since if $\alpha = 1$ the landlord does not have any incentive to provide the management input (see equation 4.9). In the case of the fixed-wage contract, the worker has no incentive to work (see equation 4.8). Thus the share contract is most likely to be chosen in equilibrium.

The optimum L can be solved as a function of α, β, and m in equation 4.8, and the optimum m as a function of α, β and L in equation 4.9 on the assumption of Nash behaviour. The optimum contract is derived from the maximization of the landlord's utility with respect to α and β, subject to the reaction function of L and m and the reservation-utility constraint $U = V$.

Note that the landlord may choose the fixed-rent contract in which the worker provides management input as well as work-effort, or owner-farming, in which the landlord provides both of these inputs if he can attain higher utility from one of these contracts than from the share contract. To the extent that management ability plays a critical role in farming, their model may help to explain some stylized facts, e.g. the dominance of owner-farming by small landholders who might have better management ability than landless farm-workers (see Section 1.2) and the prevalence of fixed-rent contracts among absentee landlords and widows who would not be adept at farm management (see Sections 6.3 and 7.2). Their share-tenancy equilibrium, however, may lose relevance if the contract is long-term.

First, let us examine why the fixed-rent contract is not chosen in their share-tenancy model despite the fact that the landlord can obtain the higher utility if he irrevocably offers in advance such a contract and refrains from shirking in accordance with equation 4.9; that is, if he maximizes u with respect to α, β, and m subject only to the reaction function of L, in addition to the reservation-utility constraint, $\alpha = 1$ is obtained, in which the landlord provides the socially optimum m and obtains higher utility than before. The problem is that, once the fixed-rent contract is accepted by the tenant, the landlord can obtain still higher utility by breaking his promise to provide the first-best m under the fixed-rent contract. In other words, such a fixed-rent contract in the one-period model does not satisfy the incentive compatibility condition that ensures the honest behaviour of the landlord. The situation is typical of a prisoner's dilemma in the absence of punishment of dishonest behaviour.[13]

For the same reason, the fixed-wage contract does not satisfy the incentive compatibility condition for the worker in the one-period model unless a sufficiently heavy penalty is imposed on a breach of contract because it is always advantageous for him to supply zero effort under the fixed-wage contract regardless of his promised effort-level. However, as we have seen in the previous section, the fixed-wage contract can satisfy the incentive compatibility condition in the long-term contract model because of the penalty on breach of contract in the form of reduced future welfare for the worker.

In the same vein, if the contract is long-term and reputation plays a role, the fixed-rent contract may become feasible in Eswaran and Kotwal's (1985b) model of tenancy contract too. To show this, assume that the landlord obtains the highest utility u_H when he offers the fixed-rent contract and breaks his promise to provide the appropriate m. Then, as a lower output will be realized and the worker's reservation utility constraint will be violated, the landlord's failure to keep his work may be revealed to other workers. In subsequent periods, no worker will accept the fixed-rent contract, and the share contract may become the best feasible option for the landlord, under which he acquires the utility of u_S. Now assume that if he offers the fixed-rent contract and behaves honestly, he will obtain the utility of u_F, which is lower than u_H but higher than u_S. The incentive compatibility condition that ensures the landlord's honest behaviour is

$$u_H + (\delta/(1 - \delta))u_S \leq (1/(1 - \delta))u_F \qquad (4.10)$$

The one-period model corresponds to the case with a zero discount factor, in which the incentive compatibility condition requires $u_H \leq u_f$. This inequality cannot be satisfied, implying that the landlord behaves dishonestly in the one-period contract. As δ becomes larger, the future utility becomes more important and the incentive compatibility condition is more likely to be satisfied. To put it differently, to the extent that the landlord takes account of his future reputation and that the penalty for dishonest behaviour in the form of reduced future utility is important, the fixed-rent contract, in which the landlord provides management input, tends to dominate the share contract in the management-effort trade-off model of Eswaran and Kotwal.

The same theory applies to the trade-off between the tenant's work-effort and his abuse of land resources (the land-quality transaction model of Section 3.2). As formulated by Murrell (1983) and others, under the one-period contract the tenant will attempt to maximize his utility with no regard to the depletion of soil fertility and other damages to the land which will adversely affect the productivity of land in the future. This tendency is likely to be stronger under the fixed-rent than the share contract because of the higher income gains from abuses of land under the former as illustrated in Fig. 3.3. Depletion of soil fertility and damage to land texture, however, will become evident over time, as output in the tenanted land declines. Then the tenant will not only be likely to be evicted but will also reduce his opportunities to lease land from other landlords. Therefore, if the contract is long-term and the tenant is concerned with his future reputation, he is more likely to have respect for land conservation in addition to applying adequate work-effort for current production.[14]

Thus, reputation, if it is recognized in the community, is likely to improve the efficiency of a contract and affect the choice of a contract in the context of long-term contracts. Of course, reputation does not always solve the problems associated with inadequate work-effort, management input, and care of the land. However, in the context of an enduring contractual relationship between landlord and worker (tenant or permanent labourer) in a relatively closed village society, reputation should play a significant role in enforcing the terms of contract, even when dishonest behaviour by the contracting parties cannot be readily verified in the short run. In other words, the long-term contract will bring an inefficient outcome under the unenforceable contract closer to the first-best situation.

Notes to Chapter 4

1. The assumption of zero discount rate in their models is crucial for this result, since the penalty in the distant future will be effective in such a situation.
2. This section partly draws on Otsuka *et al.* (1993).
3. Our model, however, is more general than theirs in the specification of utility and production functions.
4. It is empirically known that the tasks performed by permanent labourers are largely the same as tenants as well as owner-farmers (Hirashima 1978; Otsuka *et al.* 1989). See also Ch. 9.
5. We do not assume the imposition of an explicit penalty on the shirking of the worker in the form of cash payment because such a penalty is seldom observed in practice. The implication of our model remains unchanged even if we introduce such a penalty payment.
6. Labour-contract models of Lucas (1979), Feder (1985), and Eswaran and Kotwal (1986) commonly assume that the labourer's work-effort is an increasing function of the landowner's supervision time and that the wage is paid in proportion to the labourer's work-effort. As Singh (1989) correctly points out, however, their specification of the role of supervision is untenable because the work-effort should depend on the rewards and penalties attached to the results of supervision. Moreover, so far as work-effort is inaccurately monitored, the wage payment cannot be made to depend on it directly. To be consistent with the assumption of imperfect monitoring, mode of wage payment must be a fixed-wage independent of the work-effort.
7. The assumption of an infinite horizon is justifiable because the outcomes yielded by such a model are known to be more realistic than those by the finite horizon counterpart (Benoit and Krishna 1985). Furthermore, it could be replaced by the weaker assumption that the horizon is finite but sufficiently long (Benoit and Krishna 1985) or that the horizon is finite but uncertain (Luce and Raiffa 1957; Yaari 1965).
8. See Otsuka *et al.* (1993) for details of the formulation of the optimization problem and the optimum conditions.
9. A similar result has been obtained by the efficiency-wage models initiated by Shapiro and Stiglitz (1984), which also assume costly monitoring of workers' effort and the payment of fixed wages.
10. Uncertain wage in the spot casual-labour market can be incorporated into our permanent-labour contract model by considering that W_c in equation 4.2 is uncertain.
11. At the outset E-K assume that the landlord faces a choice of one of three alternative contracts: (*a*) the fixed-wage contract in which the landlord provides management input and supervision time of casual

labourers, (*b*) the fixed-rent contract in which the farm-worker provides both of these inputs, and (*c*) the share tenancy contract in which the landlord provides management input and the worker provides supervision time. They argue that the landlord tends to prefer the fixed-wage (fixed-rent) contract to the share contract when he is endowed with higher (lower) abilities of supervision and management than the worker. The choice of share contract presupposes that the landlord has superior ability of management and the worker has superior ability of supervision. However, the assumption that the landlord or the worker provides the supervision input of casual labourers does not seem to be valid because the supervision of casual labour is relatively easy, as E-K's permanent-labour contract model assumed as well as we argued in Ch. 1. Therefore we reinterpret the worker's supervision input in E-K as the work-effort here. Corresponding to this interpretation, the fixed-wage contract in E-K can be reinterpreted as owner-farming, since the landlord is considered to provide his own work-effort instead of supervising casual labourers.

12. In their original model, the cost of effort is measured by the prevailing market wage-rate per L; thus tenant's utility is replaced by his income net of the forgone wage-earning. And a similar assumption is employed for the landlord.
13. For a survey of recent game-theoretic attempts to explore the conditions under which this dilemma can be resolved, see Wilson (1985).
14. Indeed, the trade-off models of the Murrell and the E-K varieties are formally analogous. Fig. 3.3 can be interpreted as representing the Eswaran and Kotwal (1985*b*) model of tenancy contract if the work-effort and land-quality monitoring curves are relabelled as the curves of efficiency losses due to shirking in the tenant's work-effort and the landlord's management input. The same as for the Murrell model, the equilibrium of the E-K model is not Pareto-optimal so long as both the landlord and the tenant take the Cournot–Nash behaviour of 'followers'.

5

Interlinked Contracts

A LANDLORD and his tenants, permanent labourers as well, simultaneously enter into several contracts—a situation known as interlinked contracts in the literature on agrarian institutions (Bardhan 1980; Hart 1986; Bell 1988). Typically the landlord provides consumption credit to his tenants and permanent labourers at an interest rate lower than the prevailing market rate or even free of any charge (Gapud 1959; Thorner and Thorner 1962; Breman 1974; Rudra 1975; Mangahas et al. 1976; Finkler 1978; Husken 1979; Bardhan and Rudra 1980, 1981; James and Roumasset 1984; Jodha 1984; Nabi 1986). It is also common for the landlord to give his tenant/labourer grants for emergency relief on such occasions as the sickness or death of a family member (Breman 1974; J. C. Scott 1976; Hayami and Kikuchi 1982). Further, the landlord often shares the cost of purchased inputs, such as fertilizer and chemicals with his share tenants, almost universally at the same rate at which output is shared (Schickele 1952; Parthasarathy and Prasad 1974; Castillo 1975; Shlomowitz 1979; Binswanger et al. 1984; Bell and Srinivasan 1985a; Nabi 1986). In this cost-sharing arrangement, the landlord usually provides the purchased inputs at the time of their application and deducts the amount of output corresponding to the input cost before apportioning output (Gapud 1959; Rudra 1975). The provision of inputs, therefore, involves an element of *de facto* production loans to tenants.

As the land is the most suitable and important asset for collateral in less developed agrarian economies (Binswanger and Rosenzweig 1986; Binswanger and McIntire 1987), the landlord is likely to have better access to cheaper credit sources than landless tenants and labourers.[1] Therefore, interlinked contracts are likely to be a response to imperfect credit markets. Classical economists, Marshall (1890), and Schultz (1940) recognized the role of share tenancy in the world characterized by the capital-market imperfection.[2] Important questions are why the interest rate charged by landlords is usually so low and why output and input costs are shared equally between landlord and share tenant.

Following an excellent summary of the issues by Bardhan (1980),

Interlinked Contracts

various theories have emerged to explain the prevalence of interlinked contracts. Their results, however, are vastly different, although they are not necessarily mutually conflicting. For example, while Braverman and Srinivasan (1981) identify the imperfection of credit markets as a major reason for the interlinking of share tenancy and credit contracts, Braverman and Stiglitz (1982) and Mitra (1983) emphasize the incentive effect of a tenant's borrowing on his unenforceable work-effort. With regard to cost-sharing, although the effect on the share-tenant of the incentive of subsidizing the cost of purchased inputs through input complementarity is commonly identified as its major underlying motive, the ways in which its choice is justified are markedly different among different theories.

In this chapter we extend the basic model developed in Chapters 2 and 3. We show, first, that the phenomenon of interlinked credit contracts is closely related to imperfections in credit markets; secondly, that different theoretical results in the literature on interlinked credit stem mainly from different assumptions regarding the enforceability of the farm-worker's borrowing from outside sources and purchased inputs by the landlord; thirdly, that consumption credit and shared costs play essentially the same role as the fixed payment β in the basic model; and finally, that leading theories of interlinked contracts adopt assumptions too restrictive to be generally acceptable, but they do provide clues for understanding the puzzle of low interest rates in interlinked credit contracts, the dominance of 50:50 share tenancy, and equal output and input cost-sharing rules.

5.1 Interlinked Credit Contracts

In this section we will show first a general model of interlinked credit contracts and then characterize the main features of leading theories of interlinked contracts in terms of the assumptions on collateral requirement of credit. Following Braverman and Stiglitz (1982), we assume a crop season consists of two periods, e.g. planting and harvesting. The landless farm-worker receives income only in the second period, so that he has to finance first-period consumption by credit. The worker's production function is redefined as

$$Q = \theta F(L_1, L_2), \tag{5.1}$$

where L_i refers to worker's-effort in period i ($i = 1, 2$). The worker maximizes the expected discounted sum of utilities over two periods:

$$E\bar{U} = U(C_1, L_1) + \delta EU(C_2, L_2), \tag{5.2}$$

where C_i denotes consumption in period i ($i = 1, 2$) and δ is a utility discount factor. In equation 5.2, all variables except for C_2 are assumed to be determined before state θ is actually observed. Assuming credit is available both at the regular market and through interlinked contracts, the worker's *ex ante* intertemporal budget constraint is

$$\begin{aligned} C_1 + C_2/(1 + r_T) &= \{\alpha\theta F(L_1, L_2) + \beta\}/(1 + r_T) \\ &\quad + \{B_L + B_T\} - \{(1 + r_{LT}) \\ &\quad B_L + (1 + r_T) B_T\}/(1 + r_T) \\ &= \{\alpha\theta F(L_1, L_2) + \beta + (r_T - r_{LT}) \\ &\quad B_L\}/(1 + r_T), \end{aligned} \tag{5.3}$$

where r_T stands for the exogenously given interest rate on credit available to the worker on regular credit markets, r_{LT} stands for the interest rate determined by the landlord for the interlinked credit, and B_T and B_L represent the amount of credit borrowed from the regular credit market and from the landlord, respectively. We assume here that both B_T and B_L are determined before state θ is realized. Following the intertemporal utility-maximization models of Yaari (1965: 139), Arrow and Kurz (1970: 158), and Moffet (1978), and the one-period interlinked share-tenancy model of Braverman and Srinivasan (1981), we also assume that the worker can borrow B_L at r_{LT} from the landlord and B_T at r_T in the regular markets using his expected income at the time of harvest as collateral, i.e.

$$B_L + B_T \leq \frac{\alpha F(L_1, L_2) + \beta}{1 + (1 - \eta)r_T + \eta r_{LT}}, \tag{5.4}$$

where $\eta(B_L/(B_L + B_T))$ is a proportion of the worker's borrowing from the landlord. It is evident from equations 5.3 and 5.4 that, if $r_T < r_{LT}$, no voluntary interlinking occurs; if $r_T = r_{LT}$, the credit from the landlord in no way affects the worker's budget constraint; and if $r_T > r_{LT}$, the worker borrows from the landlord as much as possible.

However, the uncertain harvested crop (or the worker's share plus his fixed payment) in equation 5.4 may not be acceptable collateral especially for the outside creditors. This is because the risk of default arising from production uncertainty cannot be perfectly insured due to moral hazard (the failure of the borrower to observe the socially efficient contractual terms) and/or adverse selection problems, as intensively analysed in the theories of credit rationing (Stiglitz and Weiss 1981; Bester 1985; Stiglitz 1987).[3] If so, the amount of borrowing from outside creditors might be constrained to be less than desired at r_T. In the extreme form of rationing where the worker's future income does not command any collateral value, r_T becomes infinite and hence equation 5.3 is rewritten as

$$C_1 = B_L \tag{5.5a}$$
$$C_2 = \alpha\theta F(L_1, L_2) + \beta - (1 + r_{LT}) C_1. \tag{5.5b}$$

Furthermore, even for the landlord, the collateral value of the worker's uncertain future income might be less than its expected value. In such a case, equation 5.4 must be reformulated as

$$B_L \leq [\alpha F(L_1, L_2) + \beta]/(1 + r_{LT}). \tag{5.6}$$

In this rationing case, the worker's first-period consumption decision is identical to his borrowing decision. And once equation 5.6 becomes binding, both consumption and work-effort decisions are controlled by the landlord via a change in B_L.

Leading models of interlinked-credit contract by Braverman and Stiglitz (1982) and Mitra (1983) assume without any justification that the maximum amount of credit provided by the landlord is independent of the worker's future income and that workers have no opportunity to borrow from outside creditors. In this case, the worker's borrowing constraint may be expressed as

$$B_L = B_{\max}. \tag{5.7}$$

With this assumption, first period consumption C_1 must be equal to the maximum credit ($B_{\max}$). As a result, the worker is allowed neither to consume more than nor less than $B_{\max}$ in period 1. In other words, the worker's consumption is perfectly enforceable by the landlord. Even the extreme form of rationing expressed in equations 5.5 and 5.6, however, does not always justify $C_1 = B_{\max}$, because the worker may want to consume less than $B_{\max}$ in period 1.

While it is true that it is advantageous for the landlord to control C_1 so long as C_1 is enforceable (Mitra 1983), it is still not clear why $B_{\max}$ must necessarily be independent of the worker's future income. Furthermore, the repayment is assumed to be made with certainty in these models, despite the presence of uncertainty in production.

Claiming the assumption of $C_1 = B_{\max}$ to be too restrictive and contradictory to observation in India where tenants/labourers generally participate in the regular credit markets, Datta et al. (1988) propose a model in which the worker (or the labourer in their model) is allowed to borrow credit at r_T. Their model, however, arbitrarily assumes $r_{LT} = 0$. Basu (1983, 1987) and Gangopadhay and Sengupta (1986) also assume $C_1 = B_{\max}$ but allow the tenant to choose $B_{\max}$. This specification is self-contradictory because the assumption of $C_1 = B_{\max}$ implies that the landlord controls the worker's consumption, who would prefer to choose $B_{\max}$ in so far as $B_{\max}$ systematically affects his work-effort. Earlier models of interlinked contracts by Bhaduri (1973) and Bardhan (1977, 1979a) regard r_{LT} as exogenously given. Due to the restrictive assumptions employed, these models commonly fail to provide a sound justification for the existence of interlinked contracts and, hence, will not be discussed here further.

Under the neo-Marshallian assumption of unenforceable work-effort under uncertainty, the problem in the models of Braverman and Stiglitz (1982) and Mitra (1983) amounts to the landlord's choice of α, β, C_1, and r_{LT}, given the worker's choice of L_1 and L_2. The worker's maximization problem is formulated as

$$\max_{\{L_1,L_2\}} E\tilde{U} = U(C_1, L_1) + \delta EU[\alpha\theta F(L_1, L_2) + \beta - (1 + r_{LT})C_1, L_2]. \tag{5.9}$$

It is obvious that α and β play essentially the same roles in this model as in the basic model of Chapters 2 and 3, while the repayment, $-(1 + r_{LT})C_1$, appears as a new element in the utility function in period 2. The landlord, assumed to be risk-neutral and non-cultivator here, maximizes his expected income subject to the reservation-utility constraint, $E\tilde{U} = V$, and the reaction functions of the worker's effort:

$$\max_{\{\alpha,\beta,C_1,r_{LT}\}} y = (1 - \alpha)F(L_1, L_2) - \beta + (r_{LT} - r_L)C_1, \tag{5.10}$$

where r_L is the exogenously given opportunity cost of capital for the landlord.

It can be easily shown from equations 5.9 and 5.10 that to determine the worker's fixed revenue in period 2, $\beta - (1 + r_{LT})C_1$, the landlord needs to manipulate either β or r_{LT} for his optimization, but not both variables. Braverman and Stiglitz (1982) assume the case of pure share-cropping, i.e. $\beta = 0$, while Mitra (1983) assumes the perfectly competitive credit market which assures $r_{LT} = r_L$. Thus, these two models have essentially the same structure though the former model is more realistic because explicit fixed payments are seldom observed. These models cannot explain the prevalence of low interest rates charged by landlords because r_{LT} may or may not be greater than r_T in the former and $r_{LT} = r_L$ is assumed in the latter.

Analogous to the effect of β on L in our basic model in Chapter 2, C_1 is shown to affect L_1 and L_2 in general. If $\partial L_i/\partial C_1 > 0$, the landlord requires the worker to consume more than he would if he could choose C_1 freely.[4] It is to this effect that both Braverman and Stiglitz (1982) and Mitra (1983) attribute the advantage of interlinked over delinked tenancy contracts. Aside from the complications introduced by the nature of the two-period model, however, their models are not fundamentally different from our basic model. Consequently, these theories do not add more insight than the inclusion of β in our basic model.

In India, tenants borrow credits from diverse sources (Bell 1988; Bell and Srinivasan 1989) which suggests that the landlord does not perfectly control the tenant's consumption. Moreover, the fact that the empirically observed value of r_{LT} is consistently lower than r_T also suggests the unenforceability of credit transactions (see equations 5.3 and 5.4).

Assuming that various sources of credit are available for the worker, and that the collateral constraint in equation 5.4 holds with an equality in the absence of production uncertainty, Braverman and Srinivasan (1981) 'explain' the phenomenon of low values of r_{LT} by showing that r_{LT} is indeterminate. The landlord's expected income now can be shown as

$$y = (1 - \alpha) F(L_1, L_2) - \beta + (r_{LT} + r_L)B_L$$
$$= F(L_1, L_2) - [\alpha F(L_1, L_2) + \beta]\frac{[1 + (1 - \eta)r_T + \eta r_L]}{1 + (1 - \eta)r_T + \eta r_{LT}}. \quad (5.11)$$

It is obvious from this equation that, given α, β, and r_{LT}, the landlord's income is larger, the smaller is $(1 - \eta)r_T + \eta r_L$. Thus, if $r_T > r_L$, the landlord has an incentive to require the worker to borrow only from him, which ensures $\eta = 1$. If $r_T < r_L$ holds, however, the landlord will not offer the interlinked credit contracts. Thus, Braverman and Srinivasan conclude that the interlinked credit contracts come into effect as an institutional response to the imperfect credit market manifested in $r_L < r_T$.[5]

If $\eta = 1$ holds under certainty, equation 5.4 is rewritten as

$$B_L = \frac{\alpha F(L_1, L_2) + \beta}{1 + r_{LT}} \quad (5.12)$$

and equation 5.11 as

$$y = \{1 - \alpha[(1 + r_L)/(1 + r_{LT})]\}F(L_1, L_2) \\ - [\beta(1 + r_L)/(1 + r_{LT})]. \quad (5.13)$$

Substitution of 5.12 into equation 5.3 results in

$$C_1 + C_2/(1 + r_T) = \{[\alpha(1 + r_L)/(1 + r_{LT})]F(L_1, L_2) \\ + [\beta(1 + r_L)/(1 + r_{LT})]\}/(1 + r_L). (5.14)$$

It is clear from equations 5.13 and 5.14 that the contracting parties are interested only in $\alpha'(\equiv \alpha(1 + r_L)/(1 + r_{LT}))$ and $\beta'(\equiv \beta(1 + r_L)/(1 + r_{LT}))$, in which a lower value of r_{LT} can be exactly offset by higher values α and β. Therefore, r_{LT} can be arbitrarily chosen. Thus, within a framework of the Braverman and Srinivasan model, there is no conflict between the prevalence of low r_{LT} and the rational behaviour of the landlord.

The assumption of certainty by Braverman and Srinivasan, however, seems too restrictive. If production uncertainty exists, it is unreasonable to assume that the worker's expected future income is fully acceptable as collateral in the outside market.[6] Yet it may not be so unrealistic to assume that his future income can still be accepted as collateral by the landlord. The advantage of landlords as creditors to workers rests on the so-called economies of scope in contract enforcement; the supervision of the worker's effort will increase both output and the chance of repayment of the loan, and the threat to the worker to cease all the contracts once moral hazard in one contract is detected will work as a strong enforcer of all contracts. Thus, 'while a poor sharecropper may have few assets acceptable as collateral in the outside credit market, his landlord would accept the tenancy contract itself as collateral; the latter . . . is in the best position to enforce repayment. . . . at the time of harvest sharing' (Bardhan 1980: 88).

Certainly, with uncertainty, default in credit repayment by the worker may occur occasionally (Shetty 1988). As long as the tenancy contract lasts a number of years, however, it may be possible to make a contractual agreement in which the worker implicitly pays the premium in good crop years for default or insufficient repayment in bad crop years (Eswaran and Kotwal 1989). Whenever such long-term credit with implicit insurance becomes feasible, it can be shown that the equations formally identical to equations 5.13 and 5.14 may hold so that the indeterminancy of the contractual terms found in the certainty model emerges again. For example, the indeterminacy can occur if the worker is guaranteed by the landlord to receive additional income of $I(\theta)$, which is either positive or negative, depending on the expected and the realized output in the following manner:

$$I(\theta) = \{\alpha F(L_1,L_2) + \beta - [\alpha\theta F(L_1,L_2)\,\beta]\}$$
$$= (1 - \theta)\,\alpha F(L_1,L_2), \qquad (5.15)$$

which in effect means that his land rent is reduced in years of poor crops (i.e. when $\theta < 1$) in return for the additional land rent in years of good crop (i.e. when $\theta > 1$). It is interesting to note here that in practice the leasehold rent (usually referred to as 'fixed' rent) is reduced in years of poor crops, as reported by a large number of case-studies (Hendry 1960; Berry 1962; Cheung 1969; Ho 1976; Mangahas et al. 1976; Horii 1981; Chao 1983; Fujimoto 1983). This suggests that the long-term contract serves the role of insurance substitute and that the leasehold contract with a rent-reduction clause may be chosen, even if the worker is risk-averse.

With such an additional insurance arrangement, the *ex ante* budget constraint in equation 5.3 becomes identical to equation 5.14. The 'default' occurs in the short run, but 'repayments' are made in subsequent periods. In other words, if such an actuarially fair insurance as shown in equation 5.15 is provided by the landlord, the risk of default by the worker disappears and, hence, equations 5.13 and 5.14 will always hold. Strategic default, however, may occur, in which the worker promises to work hard and borrows credit as much as possible in advance but actually shirks and defaults. Bell (1988) considers such a case in his one-crop season model of interlinked share-tenancy contracts. The possibility of strategic default, however, can be reduced to an insignificant level because, as Myerson (1979) and Holmstrom and Myerson (1983) have shown, the contract with incentives to tell the truth dominates other contracts without such incentives under the assumed information asymmetry. If the contract is long term and the worker loses his reputation when his shirking is detected, as in the model of the long-term contract specified in Chapter 4, the worker would have an incentive to observe the contract terms on the minimum work-effort and repayments agreed upon *ex ante*. Then the long-term contract could be chosen with the effect of reducing the incidence of strategic default.

We do not assert that the credit constraint expressed in equation 5.12 and the budget constraint in 5.14 are realistic. The point we emphasize here is that if the amount of consumption credit is related to the worker's future income, the determination of one of the three variables, α, r_{LT}, and β will be left to the discretion of the landlord, who is interested only in the 'effective' contractual parameters, like α' and β'. Such an indeterminacy suggests that interlinked contracts provide part of a necessary condition to explain not only the puzzle of the low interest rates of interlinked credits but also the greater one of the prevalence of 50:50 sharing. However, they are not sufficient to account for these puzzles. For this it must be assumed that the 50:50 share has traditionally served in a community as 'a golden rule' of justice (Murrell 1983) and its alteration by a landlord is expected to entail social opprobrium. If so, he would prefer adjustment in β and r_{LT} to that of α in response to possible changes in production technology and market conditions.

5.2 Cost-Sharing Contracts

Traditionally the advocates of the Marshallian theory of share tenancy considered that, under certainty, the cost-sharing arrangement by which input costs are shared between the landlord and the tenant at the same rate as output results in the optimum resource-allocation (Schickele 1941; Heady 1947; Castle 1952; Issawi 1957;

Adams and Rask 1969; C. H. H. Rao 1975). The idea is that the share tenant's disincentive to apply inputs under output-sharing can be exactly offset by the subsidy to the input costs under the equal output- and cost-sharing rule. This view is still widely held (e.g. Currie 1981; Nabi 1986), but, as we shall see, it is invalid.

Under the Marshallian assumption, cost-sharing cannot be applied to the tenant's labour input or work-effort because by assumption such input is unobservable by the landlord. Therefore, earlier Berry (1962) and later Newbery (1975a), Roumasset (1976), A. Sen (1981), Bliss and Stern (1982), Jaynes (1982, 1984) consider that cost-sharing can occur only for those inputs that are observable and, hence, enforceable. However, a more recent study by Bardhan and Singh (1987) considers the possibility of cost-sharing for unenforceable inputs. According to their model, the income of the share tenant is defined as

$$Y = \alpha\theta F(L, x) + \beta - \gamma p x^* + p(x^* - x), \quad (5.16)$$

where x and x^* represent the amount of purchased input, say fertilizer, used by the tenant for production and provided by the landlord, respectively; and p corresponds to its real market price; γ is the tenant's share of fertilizer cost. The last term shows the gain of the tenant arising from reselling the fertilizer at the market. The question which immediately arises is why the landlord does not simply raise β rather than take the trouble of providing fertilizer, knowing that it will be resold by the tenant. If reselling entails some transaction cost, as these authors also assume later on, it is obviously advantageous for both the landlord and the tenant to change β, while leaving the decision of fertilizer purchase entirely to the tenant. Therefore, it is more reasonable to assume that the cost-sharing occurs only when the purchased input is enforceable.

The tenant's income under the cost-sharing contract for the enforceable input can be shown as

$$Y = \alpha\theta F(L, x) + \beta - \gamma p x. \quad (5.17)$$

Now suppose that in accordance with the Marshallian line of reasoning, the tenant chooses x as well as L to maximize his expected utility. The landlord, knowing the amount of x actually applied, shoulders $(1 - \gamma)$ per cent of px. The first-order conditions are

$$EU_1 \theta\alpha F_x - EU_1\gamma p = 0 \quad (5.18)$$
$$EU_1 \theta\alpha F_L + EU_2 = 0, \quad (5.19)$$

where F_x and F_L denote the partial derivatives of F with respect to x and L, respectively. The traditional Marshallian argument assumes $\alpha = \gamma$. Then the disincentive to apply fertilizer input disappears because α and γ are cancelled out in equation 5.18. So long as share tenancy is chosen, however, the work-incentive is still distorted, as can be seen from equation 5.19. As a result, the equal-sharing rule is not optimal for the landlord unless the tenant's work-effort is also enforceable (Roumasset 1976; Bliss and Stern 1982; Jaynes 1982, 1984).[7] As Braverman and Stiglitz (1986a) have argued, when the tenant's work-effort cannot be monitored, the optimum α, in general, differs from γ.

Therefore, first-best efficiency is not restored by the rule of equal output and input cost-sharing if the work-effort is unenforceable. Then why is the equal-sharing rule so pervasive in practice contrary to the presumption of Braverman and Stiglitz (1986a: 642) that 'there are frequent departures from the simple rule of setting the cost share equal to the output share'? In this connection, it is interesting to observe that γ becomes indeterminate in Braverman and Stiglitz's (1986a) model if x is directly determined by the landlord. In other words, their equivalence theorem asserts that corresponding to every cost-sharing contract which we analysed, there exists a contract with the same outcome by simply specifying the level of x without sharing the cost.[8]

To show this, denote the solution of equations 5.18 and 5.19 by $\hat{x} = \hat{x}(\hat{\alpha}, \hat{\beta}, \hat{\gamma})$ and $\hat{L} = \hat{L}(\hat{\alpha}, \hat{\beta}, \hat{\gamma})$. The income of the tenant without cost-sharing can be written as

$$Y = \bar{\alpha}\theta F(L,\bar{x}) + \bar{\beta} - p\bar{x} \tag{5.20}$$

where $\bar{x}$, $\bar{\alpha}$, and $\bar{\beta}$ are determined by the landlord, and only L is determined by the tenant. Now suppose that the landlord sets

$$\bar{\alpha} = \hat{\alpha}$$
$$\bar{x} = \hat{x}$$
$$\bar{\beta} = \hat{\beta} + (1 - \hat{\gamma})\, p\hat{x}.$$

It is clear that the expression for the tenant's income in equation 5.20 becomes identical to equation 5.17, which corresponds to the case of cost-sharing. Correspondingly, there is no difference in the form of the landlord's income equation between the two cases. Furthermore, the choice of L by the tenant without cost-sharing according to equation 5.19 remains unchanged, because the values of all other variables are the same as before. Thus, the equivalence result is established.

In the case without cost-sharing, the landlord implicitly subsidizes the cost of fertilizer by $(1 - \hat{\gamma})p\hat{x}$, but because it is part of the fixed payment β the value of $\hat{\gamma}$ becomes indeterminate. Therefore, if the landlord decides the amount of non-labour inputs, which is predominantly the case as reported by a large number of empirical studies (Ely and Galpin 1919; Taylor 1943; Berry 1962; Cheung 1969; Vyas 1970; Reid 1973, 1976a, 1979b; Parthasarathy and Prasad 1974; Rudra 1975; Husken 1979; Truran and Fox 1979; Bardhan and Rudra 1980; Bliss and Stern 1982; Bell and Srinivasan 1985a), the optimum value of $\hat{\gamma}$ becomes arbitrary.[9] Therefore any cost-sharing rule becomes immaterial and cost-sharing simply justifies the inclusion of a fixed payment in our basic model. Braverman and Stiglitz's argument that the cost-sharing rate is indeterminant is not inconsistent with the equal output and input cost-sharing rule but is unable to explain why this particular rule is pervasive.

An alternative perspective on the prevalence of equal-sharing rules is to regard the supplied inputs as *de facto* production loans. As mentioned earlier, the usual practice of cost-sharing is for the landlord to provide purchased inputs at the time of their application and deduct the amount of output corresponding to the cost of purchased inputs before sharing output. The question is how the value of output corresponding to the input costs is calculated. Production takes time in agriculture and prevailing market interest rates are very high in less developed agrarian economies. It is thus inconceivable that the interest cost should be disregarded in cost-sharing contracts. As a matter of fact, Fujimoto (1986) found in Java that the implicit interest rate charged by the landlord, estimated by the ratio of the imputed value of output deducted for the cost-sharing to the actual cost of purchased inputs, was as high as 50 per cent. This raises the question of the relevance of existing cost-sharing models that ignore the element of production credit in the cost-sharing arrangement. It may be more rewarding to analyse cost-sharing contracts within a multi-period framework with due consideration of the endogeneity of the implicit interest rate, parallel to the model of interlinked credit contracts.

5.3 On the Advantage of Interlinked Contracts

In this chapter we have critically reviewed the existing models of interlinked contracts. We have shown that the existing models have failed to provide satisfactory explanations for the stylized facts of agrarian contracts, such as the prevalence of 50:50 share tenancy, low interest charges on the interlinked credit, and equal sharing of output and input costs. The major shortcoming of these models is their inability to explain the choice of interlinked over delinked contracts. This choice cannot be explained by the incentive effects of the worker's receiving credit on his work-effort because the credit essentially plays the role of β in our basic model formulated in Chapters 2 and 3. This implies that the credit can be replaced by an explicit fixed payment. Similarly, the choice of cost-sharing cannot be explained by its positive incentive effect because the shared cost also amounts to a fixed payment.

The landlord's easier access to credit markets than his landless farm-worker may seem to provide a more plausible explanation for interlinking because the landlord facing a lower rate of interest can gain, at least potentially, from the interlinked credit transactions. Ray and Sengupta (1989), however, demonstrate that the imperfection of the credit market alone does not confer a clear advantage on the landlord offering interlinked contracts over a pure money-lender facing the same interest rate. The crux of their argument is that, if the money-lender possesses the same information as the landlord about the reaction function of the worker and his reservation utility under the assumption of unenforceable contract, he can extract the same amount of profit from money-lending as does the landlord. This suggests that in order to justify the choice of interlinked contracts, the landlord must have informational and other advantages over the money-lender, such as a more accurate assessment of the worker's effort.

It seems reasonable to assume that the landlord has a stronger incentive to supervise the worker's effort than the pure money-lender because the increased work-effort of the worker due to his supervision will increase not only his expected crop income but also the probability of credit repayment to the landlord. Moreover, provided that the landlord is a resident and experienced farmer in the community, the cost of supervision is likely to be smaller for him than for the money-lender who is typically non-resident and inexperienced in farming. The expected penalty on a worker's opportunistic behaviour can also be strengthened by offering interlinked consumption and production credits because once found out in one transaction the worker is likely to lose future opportunities of both land (or labour) and credit contracts. Penalties on opportunistic behaviour can be increased further by adding to the interlinkage other contracts such as insurance (e.g. landlord's assistance at worker's emergency need).

Thus interlinked contracts are considered an institutional adaptation to underdeveloped, imperfect markets in agrarian economies where the outcome of a worker's moral hazard is difficult to distinguish from poor results caused by nature and where each transaction is typically too small to enforce profitably by such formal means as a resort to the court. The prevalence of interlinked contracts in agrarian economies cannot be understood without considering the relative advantage of multiple transactions in information collection and contract enforcement through social interactions in small communities. The formal modelling of interlinked contracts incorporating these economies of scope is critically needed for better understanding of contract choice and enforcement in agrarian economies.

Notes to Chapter 5

1. Feder *et al.* (1988) have convincingly demonstrated that in Thailand, because of the difference in the collateral value of land, landholders possessing the legal titles of their land have greater access to formal, subsidized credit markets than those without legal titles.
2. See Cheung (1969), Bliss and Stern (1982), and Jayres (1984) for surveys of classical thought on share tenancy.
3. See Binswanger and Siller (1983), Binswanger and Rosenzweig (1986), Braverman and Guasch (1986), and Carter (1988) for the application of general credit theories to the specific case of LDC agriculture.
4. If $U_{12} \leq 0$, as we assumed in the basic model, $\partial L_i / \partial C_1 \geq 0$ holds.
5. A recent study found no significant interlinking of share tenancy and credit contract in villages in Bangladesh, where landlords who owned small areas of land were not much wealthier than the tenants who also owned some land (Taslim 1988). Jodha (1984) also found in India that tenants who were larger owners of land provided credits to less wealthy landlords.

6. As Bell (1988) points out, however, traders often finance the cultivator using crop income as collateral.
7. Gangopadhyay and Sengupta (1987) apply this result to the case of interlinked output sale and credit contracts between farmers and traders, in which traders proportionally undervalue output price and subsidize interest costs at the same rates. Such result is anticipated because no unenforceable input is included in their model.
8. Earlier Bliss and Stern (1982: 88–9) also established this equivalence theorem but under the assumption of a contract that can specify the levels of L and x.
9. Braverman and Stiglitz (1986a) also assert that cost-sharing contracts have a decisive advantage if the tenant has better information regarding the productivity of fertilizer. Their assumption that the tenant decides the amount of fertilizer application, however, is not generally consistent with the evidence.

6

Global Survey of Empirical Evidence

WHILE small-scale owner-cultivation is the most common form of land tenure in agrarian economies under consideration, land and labour contracts are used to facilitate transfer of land and labour resources among rural households endowed with different amounts of these resources. As far as the choice of land and labour contract is concerned, our basic model discussed in Chapter 3 predicts under the assumption of unenforceable work-effort that: (a) the share contract ($0 < \alpha < 1$) will be chosen in equilibrium if production is uncertain and the farm-worker is risk-averse, (b) the fixed-wage permanent-labour contract is dominated by the share or fixed-rent contract, and (c) resource allocation under share tenancy is less efficient than under fixed-rent tenancy and owner-farming. On the other hand, the models in Chapters 4 and 5 imply that, if long-term and interlinked contracts effectively reduce transaction costs in the social environment of agrarian communities characterized by strong social interactions, the prediction of the basic model under the assumption of unenforceable work-effort may be invalidated. Therefore, the central questions for empirical research are how common permanent-labour contracts are and how significant is the inefficiency associated with share tenancy.

In this chapter we attempt to examine the factors determining the actual pattern of contract choice and contractual efficiency through a global survey of empirical evidence, even though existing empirical studies pertain mostly to Asia.

6.1 Conditions of Permanent-Labour Contracts

In the theoretical models described in previous chapters, as a risk-sharing arrangement between a risk-averse worker and a risk-neutral landlord, the fixed-wage permanent-labour contract will be chosen over the share-tenancy contract only when the worker's effort is perfectly enforceable. Given the technical difficulty of monitoring farm work, it is hard to imagine such perfect enforceability, even with strong social interactions in agrarian communities.

If so, the permanent-labour contract will be more commonly observed where land tenancy is socially or legally prohibited. This hypothesis is consistent with the observation that in the nonplantation sector in Third World agriculture, permanent labour is commonly employed in countries in South Asia such as India, Pakistan, Bangladesh, and Nepal, whereas it is rare in South-East Asian countries such as Indonesia, the Philippines, Malaysia, and Thailand.

For India, a popular argument is that the incidence of permanent labour has increased with the diffusion of modern rice and wheat varieties that induced the development of irrigation and double cropping where fertilizer and chemicals are applied heavily (Bhalla 1976; Bardhan and Rudra 1981). The introduction of a more intensive farming system increased the demand for labour, especially in the peak season and for the tasks requiring care and judgement that can be performed better by permanent labourers than by casual labourers (in the logic discussed in Section 4.2). While this argument may be valid for the choice between casual- and permanent-labour contracts in India, it fails to explain why permanent labour is rarely observed in South-East Asia including well-irrigated Java in Indonesia, where both the diffusion of modern crop varieties and increase in cropping intensity have been no less significant than in the Punjab and Haryana. In Java, the incidence of tenancy increased instead of permanent labour with the diffusion of modern rice varieties (Husken 1979; Hayami and Kikuchi 1982).

In fact, the incidence of permanent labour had been widespread in India since long before the advent of modern cereal technology or even before Independence (Thorner and Thorner 1962; Breman 1974; George 1987). In India, Pakistan, and Nepal, the permanent-labour contract is closely associated with caste status: permanent labourers belong to the lower castes and their employers belong to the higher castes (Thorner and Thorner 1962; Sanghavi 1969; Breman 1974; Hirashima 1978; Bardhan 1983; Basant 1984; Binswanger *et al.* 1984; Kotwal 1986; George 1987). Traditionally, 'the lower castes were prevented from owning or leasing land by powerful social sanction' (George 1987: 142). Where violators of the caste code are socially punished (Breman 1974; Akerlof 1976), the permanent-labour contract is expected to emerge as a substitute for the land-tenancy contract.

In contrast, countries in South-East Asia in general have no caste system comparable with that in India. Village communities in South-East Asia are relatively homogeneous socially, if not economically. Many landless labourers in Indonesia, Malaysia, the Philippines, and Thailand are relatives, even sons of peasant-cultivators, so that there are no class barriers to becoming tenant- or owner-cultivators (Hirashima 1977; Kuchiba and Bauzon 1979; Kuchiba et al. 1979; Hayami and Kikuchi 1982; Fujimoto 1983). In such an environment, the land-tenancy contract predominates and permanent-labour contracts are seldom observed.

Land-reform programmes have widened this difference in agrarian organization between South and South-East Asia. In the early post-independence period, redistributive land reform was popular in countries in South Asia. According to land-reform laws, commonly called 'land-to-the-tiller legislation', in India, Pakistan, Bangladesh, and Nepal, large holdings above a certain land-retention limit were confiscated from landlords for distribution to landless labourers, with compensation lower than market land values. This land-transfer programme applied only to tenant-cultivated land, whereas land under 'personal cultivation' by landlords using hired labour was usually exempted from the land-transfer programmes (Khusro 1969; Dantwala and Shah 1971; Appu 1975; Ladejinsky 1977; Jannuzi and Peach 1980; Herring 1983; Sharma 1987). This provision in the land-reform laws has had the effect of precluding land tenancy from the options of contract choice for landlords. Many landlords evicted tenants and converted them to permanent labourers. Khusro (1969: 140) characterized the permanent-labour contracts in India 'as a means of dodging the provision of land reform'. In general, in the Indian subcontinent land-reform legislation has contributed to significant increases in the incidence of permanent labour as a substitute for land tenancy, although its effect varies from place to place.[1] According to the National Sample Survey, the incidence of farm area under tenancy at the all-India level declined from 20.3 per cent in 1953/4 to 11.6 per cent in 1970/1, while the incidence of farms employing permanent labour increased from 13.6 per cent in 1953/4 to 19.0 per cent in 1970/1.[2]

This process is well illustrated by a history of a village in the rice-growing area of Tamil Nadu, which was covered by intensive surveys five times from 1916 to 1983 (Guhan and Bharathan 1984).

In 1916, over 70 per cent of cultivatable land in this village was under tenancy, of which over 80 per cent was share-cropped. The ratio of tenanted land remained about the same in 1937 but the incidence of share tenancy declined and that of fixed-rent tenancy increased to occupy more than 60 per cent of tenanted land, corresponding to outmigration of landowning Brahmins from the village to urban areas. In 1959 after enactment of the land-to-the-tiller legislation, it was found that the ratio of tenanted land decreased sharply to 25 per cent, which went down further to 23 per cent in 1983 with complete disappearance of fixed-rent tenancy. The share-cropping contract (*al-varum*) which has become dominant in recent years was more of a permanent-labour than a tenancy contract by which the worker (*varamdar*) supplies labour alone for the share of output at only 27.5 per cent while the landlord takes care of all other costs including the wages of casual labour employed in peak seasons.

In the market economies of South-East Asia (excluding the socialist bloc in Indo-China), land reform has been planned and implemented much less intensively. A major exception is the Philippines where since the American colonial regime land reform has been considered necessary to suppress agrarian unrest resulting from the highly skewed distribution of land assets, a legacy of the Spanish regime. Reform programmes based on the 1963 Agricultural Land Reform Code were implemented vigorously under the Martial Law proclaimed in 1972. These programmes were successful in abolishing share tenancy, fixing land rent at a low level, and imposing ceilings on landholdings mainly in advanced rice-growing areas (Herdt 1987; Otsuka 1991). The land-reform beneficiaries who were converted from the status of share-tenants to lease-holders and amortizing owners were granted the usufruct right on the redistributed land but not the right to transfer it except to legitimate heirs. A response to these land-reform regulations highly relevant to the issues we are concerned with here is the emergence of the permanent-labour contract akin to that practised in India, as a substitute for illegal subleasing of land in irrigated areas of Central Luzon. In fact, a new form of permanent labour becoming dominant in these areas resembles the *al-varum* contract in Tamil Nadu village. This 'Indianization' of the Philippine rice bowl represents a logical adjustment to limitation imposed by legislation on the scope of contract choice (see Chapter 9 for a more rigorous analysis).

Global Survey of Empirical Evidence

The cross-sectional contrast between South and South-East Asia finds a counterpart in the agrarian history of Japan. In Japan, permanent labour was preponderant in the early Tokugawa period (sixteenth to seventeenth centuries) under agrarian laws that commanded direct attachment of all cultivators to feudal lords and hence prohibited landlord–tenant relations to develop by legal peasants (*honbyakusho*) who were granted usufruct rights on land. This rule was gradually breached by *de facto* tenancy contracts between the legal peasants and their permanent labourers. The conversion of permanent labourers to tenants was accelerated in the later Tokugawa period (eighteenth to nineteenth centuries) when the feudal lords became less eager to enforce the tenancy regulation as land taxation shifted from the system of a variable levy based on crop yield assessments (*kemi*) to the fixed levy in kind (*jomen*). The conversion progressed further in the modern era as the Meiji government (1868–1912) granted modern property rights in land, including the right of renting it out, to those who used to have feudal usufruct rights on the land. This process was supported by the development and diffusion of intensive cropping systems that required more intensive care and judgement of farm-workers and, therefore, could be operated more efficiently by small tenant-cultivators who could claim residual profits (Oishi 1958; T. C. Smith 1959).

Both the cross-regional and the historical comparisons are consistent with the hypothesis that the enforceability of a fixed-wage labourer's work-effort is usually much less than perfect so that the land-tenancy contract tends to prevail over the permanent-labour contract in agrarian economies characterized by the absence of scale economies, provided that there is no institutional constraint on contract choice.

6.2 Efficiency and Income Distribution under Share Tenancy

If land-owning principals cannot perfectly enforce landless agents' work-effort, we would expect that share tenancy adopted for the sake of risk-sharing results in less efficient resource allocations than fixed-rent tenancy and owner-farming. On the other hand, tenants are expected to pay higher rents for lower risk under share than in fixed-rent tenancy. These theoretical implications as summarized in Table 3.1 will be examined here in terms of past empirical studies.

Comparison of output and input per hectare

A major problem in empirical testing of contractual efficiency is that, as shown in Table 3.1, implications for resource allocation and income distribution are exactly the same between the enforceable contract and the unenforceable contract under uncertainty if $\sigma = 1$. This means that an exact test is impossible without information on the elasticity of substitution. However, it seems reasonable to expect that if the unenforceable contract assumption really holds rather irregular relations of resource allocation and income distribution will be observed as many cases are compared for different crops and different regions that are likely to have different production parameters.

The most commonly used method of testing the hypothesis of Marshallian inefficiency under share tenancy has been to compare yields per hectare between land under share-cropping and land under owner-farming and/or fixed-rent tenancy. In order to test the significance of this Marshallian inefficiency, a large number of case-studies have been conducted, mostly in South and South-East Asia, to compare output and inputs between different tenure classes, mostly in production of rice and wheat classified by conditions of irrigation and the size of cultivation. The comparisons have been made in terms of both the physical yields of a single crop per hectare of gross sown area and the total values of output of various crops grown per hectare of net sown area. In order to make the results of these case-studies comparable, we calculate the rates of difference in output per hectare of share-cropping (Y_s) from that of owner-farming and fixed-rent tenancy (Y_o and Y_f), i.e.

$$(Y_s - Y_o) \div ((Y_s + Y_o)/2) \text{ and } (Y_s - Y_f) \div ((Y_s + Y_f)/2).$$

The rates of difference thus calculated are summarized in the form of frequency distributions in Table 6.1.

The results of comparisons between share tenancy and owner-farming in terms of single-crop output (Table 6.1, col. 1) are distributed in a smooth bell shape that can well be approximated by a normal distribution; its mean is not statistically different from

TABLE 6.1. *Frequency distribution of the rates of difference in output per hectare of share tenancy land from that of owner-farming and fixed-rent tenancy land in past case-studies* (%)

Class in the rate of difference	Share tenancy compared with		
	Owner-farming		Fixed-rent tenancy:
	Single-crop output quantity	Total output value	single-crop output quantity
−75.0 ~ −65.1	—	2.3	—
−65.0 ~ −55.1	—	2.3	—
−55.0 ~ −45.1	2.3	9.1	—
−45.0 ~ −35.1	3.7	13.6	—
−35.0 ~ −25.1	4.7	9.1	3.2
−25.0 ~ −15.1	12.1	15.9	4.8
−15.0 ~ −5.1	17.7	9.1	19.4
−5.0 ~ 4.9	23.7	20.5	43.5
5.0 ~ 14.9	15.8	13.6	16.1
15.0 ~ 24.9	8.4	2.3	8.1
25.0 ~ 34.9	5.6	2.3	—
35.0 ~ 44.9	4.2	—	4.8
45.0 ~ 54.9	0.9	—	—
55.0 ~ 64.9	0.9	—	—
TOTAL	100.0	100.0	100.0
Mean	−0.7	−16.6**	1.4
Standard deviation	21.1	23.6	13.8
Sample size	215	44	52

** Significantly different from zero at 1% level.

Sources: Ahmad (1974), Bell (1977), Berry and Cline (1979), Bhuiyan (1987), Bhuiyan and Nandal (1987), Bliss and Stern (1982), Castillo (1975), Chattopadhyay (1979), de Janvry, Fukui, and Sadoulet (1989), Dwivedi and Rudra (1973), Fujimoto (1983, 1985, 1986), Hayami and Kikuchi (1982), Heady (1955), Herdt (1978), Hossain (1977), Huang (1975), Jabbar (1977), Khandker et al. (1987), Ledesma (1982), Mandal (1980), Mangahas et al. (1976), Nabi (1986), Pal (1975), Parthasarathy (1975), Ransom and Sutch (1973), Ruttan (1966), Sen (1981), Shahid and Herdt (1982), Stewart and Arellano (1975), Talukder (1980), Tamin and Mustapha (1975), and Zaman (1973).

zero at conventional significance levels (see also upper panel in Fig. 6.1).[3] Similar results are obtained from the comparisons with fixed-rent tenancy (Table 6.1, col. 3 and lower panel in Fig. 6.1), although the distribution is less smooth because of a smaller number of sample observations. These results do not support the hypothesis that Marshallian inefficiency prevails under share tenancy.[4] Such results are also consistent with the yield function analysis in which a dummy variable representing share tenancy is found to have no significant coefficient (Chandra 1974; Roumasset 1976; Truran and Fox 1979).

These results, however, must be taken with caution. First of all, no case-study estimated the direct elasticities of substitution, so that the results of yield comparison alone do not constitute sufficient evidence to disprove the significant inefficiency of share tenancy. Secondly, large margins of error must be assumed from the case-studies as they do not often control adequately for relevant factors such as land quality and farm-worker's and owner-cultivator's abilities. However, because there is no reason to assume that these omitted factors routinely bias the results of yield comparison in one particular direction, possible biases in individual case-studies conducted in different environments under different survey designs are expected to cancel out. Thus, the law of large numbers is likely to apply. To be conservative, the statistical evidence may be taken to imply that share tenancy is not as inefficient as the Marshallian model of share tenancy assumed.

In contrast, the existence of Marshallian inefficiency is suggested from the comparisons based on total output per hectare (Table 6.1, col. 2 and central panel in Fig. 6.1) for which the mean of the rates of difference is negative and significantly different from zero at the 1 per cent level. However, the distribution is highly irregular. It appears that this irregularity stems from differences in the production function due to differences in crop mix between sharecropping and owner-farming areas. In general, there is a tendency for staple food crops, such as rice and wheat, to be grown in sharecropping areas and for high-value commercial crops to be grown in owner-farming areas (Raj 1970; Bharadwaj 1974; Kutcher and Scandizzo 1976). In fact, Heady (1955), Junankar (1976), and Bagi (1981) all found significant differences in the production function between share tendency and other tenure forms using estimates of Cobb–Douglas production functions for aggregate value of outputs.

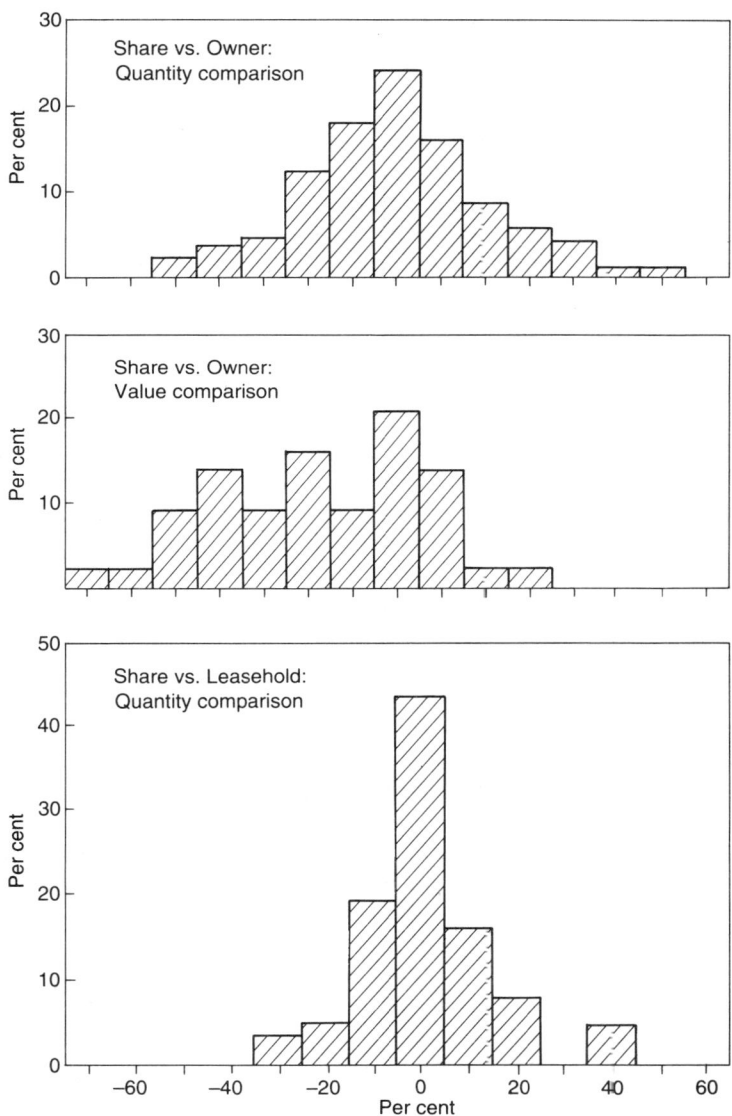

Source: Table 6.1.
FIG. 6.1 Frequency distribution of the rates of difference in output per hectare of share-tenancy land from that of owner-farming and leasehold-tenancy land

Therefore, the significantly lower average output value per hectare for share-cropping than for owner-farming areas seems to reflect more of a difference in production functions than the existence of Marshallian inefficiency which refers to suboptimal labour input per hectare for the same production function. The systematic difference in crop choice might reflect the landlord's preference, because of the relative ease of monitoring inputs, to let his share-tenants grow staple food crops characterized by a simpler and stabler production function rather than to let them grow more complicated commercial crops (C. H. H. Rao 1971; Bell 1977). If so, the difference in aggregate output may reflect a form of inefficiency although it is of a different nature from Marshallian inefficiency. This inefficiency, too, is not always insuperable; it can be made insignificant where the strong community mechanism of contract enforcement operates and where there is no institutional constraint on contract choice, as attested by the case-study in Chapter 7.

Data available from past studies for comparisons of inputs (labour and fertilizer) per hectare between share tenancy and other land-tenure forms are too small to make meaningful inferences about their distributional forms.[5] Therefore, only the means and standard deviations of the rates of difference are reported in Table 6.2.

Except for the cases of input of fertilizer per hectare under share tenancy compared with owner-farming, the results in Table 6.2 are inconsistent with the Marshallian inefficiency hypothesis. The significantly higher fertilizer input per hectare of physical land area under owner-farming is consistent with the results of a comparison of total output values shown in the second column of Table 6.1 which seem to be explained largely by the difference in crop mix between share-cropping and owner-farming land.

A major anomaly is the significantly lower fertilizer input for share-cropping than for owner-farming areas when the comparison is made in terms of gross sown area for a single crop. This result is inconsistent with the results of the single-crop output comparison (Table 6.1, col. 1) as well as with the comparison with fixed-rent tenancy in fertilizer input per hectare of gross sown area for a single crop (the last row of Table 6.2). This anomaly seems to be explained, to a large extent, by underestimation of fertilizer input by share-tenants, who tend to report only their own expenditure

Global Survey of Empirical Evidence

TABLE 6.2. *Average rates of difference in labour and fertilizer inputs per hectare of share-tenancy land from those of owner-farming and fixed-rent tenancy land in past case-studies (%)*

Rate of difference	Mean	Standard deviation	Sample size
Labour input per ha.			
vs. owner-farming:			
Single-crop output quantity	−5.1	14.5	18
Total output value	1.7	11.6	11
vs. fixed-rent tenancy:			
Single-crop output quantity	−5.7	14.3	18
Fertilizer input per ha.			
vs. owner-farming:			
Single-crop sown area	−9.9**	24.7	68
Physical land area	−12.6*	29.8	21
vs. fixed-rent tenancy:			
Single-crop sown area	2.7	15.6	11

* Significantly different from zero at 5% level.
** Significantly different from zero at 1% level.

Sources: Bagi (1981), Berry and Cline (1979), Bhuiyan and Namdal (1987), Bliss and Stern (1982), de Janvry, Fukui, and Sadoulet (1989), Dwivedi and Rudra (1973), Fujimoto (1983, 1985, 1986), Hayami and Kikuchi (1982), Heady (1955), Herdt (1978), Hossain (1977), Jabbara (1977), Mandal (1980), Mangahas *et al.* (1976), Nabi (1986), Pal (1975), Shahid and Herdt (1982), Talukder (1980), Tamin and Mustapha (1975), and Zaman (1973).

even when part of the fertilizer cost is shared by landlords (Fujimoto 1983: 118).

So far, our review of the empirical literature focused on Marshallian or neo-Marshallian inefficiency of share tenancy under the static assumption that the same production functions are used among different land tenancy contracts and owner cultivation. Dynamic inefficiency of share tenancy might arise, if share tenancy retards the speed of innovation in agriculture.

Amit Bhaduri (1973) proposes an interlinked credit model, in which the semi-feudal landlord may refuse to adopt an innovation, even though its adoption increases output. According to this model, an innovation increases the landlord's rent income but decreases his income from moneylending so as to reduce his total income. In his model, however, the landlord is not allowed to change the output sharing rate nor the interest rate after the innovation is introduced. Newbery (1975b) also considers the possibility of dynamic inefficiency of share tenancy based on the differential enforceability of the worker's effort under the old and new technologies. He argues that because of the unknown change in the structure of production function with the introduction of new technology, it becomes more difficult for the landlord to specify and enforce the tenant's work effort, and hence, the tenant is more likely to under-supply his work effort. In other words, more serious Marshallian inefficiency is expected to arise under the new technology. Thus, the landlord is more likely to refuse innovation under share contract than under fixed-rent contract. More recently Braverman and Stiglitz (1986b) examine whether an output increasing innovation necessarily increases the landlord's income within the framework of the neo-Marshallian model specified in our basic model. In their model, the worker's production function includes a technology parameter and the worker's effort reaction function also depends on this parameter along with other contract parameters. The landlord is assumed to adopt innovation, if his expected income increases. The important point is that the worker's effort may decline with the new output-increasing technology, particularly if the shift of production function is such that the expected marginal product of work effort declines. They demonstrate that there is a possibility that the landlord's income declines with the new technology because of the large decrease in the worker's effort even if he adjusts contract parameters to maximize his income.

This possibility of dynamic inefficiency, however, does not seem to be strongly supported by empirical evidence. According to a summary of the green revolution literature during the late 1960s and the early 1970s by Ruttan (1977) the incidence of share tenancy is not negatively related with the adoption of modern rice and wheat varieties. This view has been supported by various case studies conducted from the mid-1970s on the adoption of green revolution technology (Parthasarathy and Prasad 1974; Pal 1975; Mangahas *et al*. 1976; Barker and Herdt 1985; Boyce 1987; Herdt 1987; David and Otsuka 1990). More generally, a comprehensive survey of technology adoption in agriculture in developing economies by Feder *et al*. (1985) found no consistent evidence that share tenants were slower to adopt innovations. There are, however, relatively fewer empirical studies on dynamic inefficiency than on static production inefficiency. Moreover, many of these adoption studies are descriptive without statistical testing. Furthermore, as Feder *et al.* point out, econometric studies generally do not adequately control the effect of various factors on technology adoption other than the tenure status of cultivators. Thus, the evidence on dynamic inefficiency of share tenancy is not conclusive.

Comparisons of rent and tenant income

Comparisons of rent per hectare between share and fixed-rent tenancy are often unreliable because it is difficult to ascertain the extent of cost-sharing by the landlord under the share contract. Past studies, including Johnson (1950), Heady (1955), Winters (1974), and Reid (1979*b*) for the United States; Hoffman (1984) for sixteenth- and seventeenth-century France; Cheung (1969) and Chao (1983) for pre-war China; and Rao (1975), Huang (1975), Mangahas *et al.* (1976), and Fujimoto (1983) for contemporary Asia, are unanimous on the point that gross rent per hectare is lower under fixed-rent than under share tenancy.

Several studies on contemporary Asia find that net rent (gross rent minus cost shared by the landlord) is also lower by approximately 20–30 per cent under fixed-rent than under share tenancy (Parthasarathy and Prasad 1974; Umehara 1974; Castillo 1975; Herdt 1978; Hayami and Kikuchi 1982; Ledesma 1982; Tongpain and Jayasuriya 1982; Nabi 1986). Correspondingly, the income of tenants from farming per hectare is reported to be about 30 per cent lower under share tenancy than under fixed-rent tenancy (Umehara 1974; Castillo 1975; Mangahas *et al.* 1976; Herdt 1973). Similar findings are obtained from our case-studies reported in the next two chapters.

These results strongly support the hypothesis that share tenancy exists when tenants (and perhaps landlords, too) are risk-averse under the condition of uncertainty. Such results are consistent with the positive association across regions between the adoption of share tenancy and the degree of variations in yield primarily caused by the weather found in the large number of empirical studies (Cheung 1969; R. Higgs 1973; Parthasarathy and Prasad 1974; Huang 1975; Parthasarathy 1975; Bardhan 1977; Dowell 1977) with the exception of C. H. H. Rao (1971) and Reid (1973). Recently Rosenzweig (1988*a*) has provided supportive evidence that choice of share tenancy by farmers is positively related to income varaiability. Moreover, Schultz (1940) has pointed out that as farm incomes became unstable in the 1930s, crop-share leases increased rapidly at the expense of the fixed-rent cash lease in Iowa. Dowell (1977) has also consistently shown, through cross-section analysis of tenure choice in the United States from 1890 to 1970, that as stable off-farm wage-employment opportunity increases, the share contract tends to be replaced by the fixed-rent contract.

Relevance of alternative assumptions

In terms of the empirical data, which assumption is considered appropriate among the alternatives listed in Table 3.1? From the results of a comparison of land rent and tenant income per hectare, it seems safe to rule out the possibility of certainty or risk-neutrality of workers. Comparisons in output and inputs produced somewhat conflicting results, but, on balance, evidence is stronger to support the hypothesis of equal allocative efficiency between share tenancy and other land-tenure forms.

If we accept the equal efficiency hypothesis with respect to resource allocation under uncertainty, remaining alternatives are the enforceable contract or the unenforceable contract with $\sigma = 1$. It is inconceivable, however, that the elasticity of substitution is close to unity across regions and over time. Therefore, the assumption of enforceable contract is considered more relevant, even though no agricultural tenancy study has yet undertaken estimation of the elasticity of substitution in addition to comparison of resource allocation and income distribution. By taking the case of the jeepney (informal minibus in the Philippines), we will further explore the issue of tenurial efficiency based on the estimation of the elasticity of substitution between capital and work-effort in Chapter 8.

6.3 Mechanism of Effective Share-Contract Enforcement

While some cases are reported of direct supervision of share-tenants by the landlord (Alston and Higgs 1982; Alston *et al.* 1984; Nabi 1986), more often no visible supervision is found. If significant Marshallian inefficiency is not commonly detected, it is not necessarily because the share contract can be costlessly enforced in general, but because those particular landlords who adopt the share contract tend to be those equipped with a relatively efficient mechanism of contract enforcement such as the role of reputation in agrarian communities where social interactions are intense (Chapters 1 and 4). This enforcement mechanism will be stronger in more tightly structured communities in which the rights and obligations of each member are more clearly defined by tradition (Embree 1950; Hayami and Kikuchi 1982). Agricultural production technology is also a major determinant of contract enforceability. In general, the more complex and less standardized farm operations are, the more difficult it is for landlords to monitor tenants' work-effort. Within the same community under the same technology, work enforcement should be less costly for landlords who themselves are experienced farmers than for non-farmers and widows.

As an empirical proposition we postulate that if there is no institutional limitation on contract choice, landlords whose cost of enforcing worker's effort is low tend to choose share tenancy, whereas those with high-enforcement costs prefer fixed-rent tenancy even at the expense of a reduced rent that compensates tenants for greater exposure to risk. If landlords select contracts on the basis of their comparative advantage in monitoring tenants' work-effort, it is not surprising that Marshallian inefficiency is not commonly observed.

Optimizing behaviour of landlords is consistent with the common observation that small to medium-sized resident landlords tend to prefer share tenancy whereas large absentee landlords prefer fixed-rent tenancy (Takahashi 1969; R. Higgs 1974; Umehara 1974; C. H. H. Rao 1975; Bliss and Stern 1982; Hoffman 1982). Needless to say, compared with an absentee landlord, a resident landlord living together with his tenants in the same village has the advantage of monitoring his tenants' work-effort directly as well as developing a relationship of trust with his tenants. Because a landlord's personal capacity to monitor farm-work is limited, a large landowner tends to choose fixed-rent tenancy. Among relatively small landlords, the incidence of share tenancy tends to be higher when landlords are experienced in farming and when landlords and tenants are relatives or close friends (Ely and Galpin 1919; Takahashi 1969; Stewart and Arellano 1975; Finkler 1978; Robertson 1982; Cohen 1983; Fujimoto 1983, 1986; Lehmann 1986; de Janvry *et al.* 1989). This tendency is clearly indicated in our case-study of tenancy choice in West Java (Chapter 7). Personal ties developed among family members, relatives, and friends are considered highly instrumental for efficient work enforcement not only in agriculture (Hayami and Kikuchi 1982; Alston and Ferrie 1985) but also in other sectors of the economy (Ben-Porath 1980; Pollak 1985).

The hypothesis that share tenancy is adopted where the enforcement of tenant work-effort involves relatively low cost is also supported by the observation that share contracts used to apply mainly to food crops such as rice and wheat in India (Raj 1970; Bharadwaj 1974; Bharadwaj and Das 1975), for which technology was simpler and more stable, whereas fixed-rent tenancy was applied to cash crops such as cotton and tobacco, despite the greater risk involved, because more scope existed to vary tenants' inputs (C. H. H. Rao 1971).

This hypothesis should apply not only to the share contract but also to other contracts. The fixed-rent tenancy would not be adopted if it were difficult to prevent mismanagement of land and land-specific capital. This hypothesis is supported by the findings of the Philippine jeepney study which showed that a higher percentage of jeepney-owners under fixed-rental contracts are mechanics and professional drivers, both having the ability to check the damage to vehicles from reckless driving by tenants, whereas a higher percentage of owners under share contracts are relatives and friends of tenant-drivers (see Chapter 8). The fact that large farm firms based on wage labour are limited mainly to monoculture and are never successful in cases of complex crop rotation and crop–livestock mixed farming cannot be explained without resorting to the high cost involved in the enforcement of wage-labourers' work-efforts and the resultant scale diseconomies (Brewster 1950). The hypothesis of scale diseconomies arising from monitoring hired farmwork is supported by the observation that the fixed-wage time-rate contract tends to be adopted by small farms employing a small number of workers whereas the piece-rate contract tends to be adopted by larger farms (Roumasset and Uy 1980).

The choice of contract will vary as the enforcement costs associated with various forms of contract change, corresponding to changes in technology, market, and social conditions. For example, introduction of large-scale threshing-machines to Central Luzon in the Philippines facilitated a shift from fixed-rent to share tenancy

Global Survey of Empirical Evidence 101

on large estates because it reduced the cost of monitoring rice yield levels (Kikuchi and Hayami 1980; Hayami and Kikuchi 1982). On the other hand, farm mechanization, especially in cotton-picking in the US South, induced a reverse shift from share tenancy to wage labour as it reduced the number of workers involved and, also, standardized farm tasks thereby reducing the cost of monitoring (Day 1967; Alston 1981).

Thus, if landowners self-select various agrarian contracts in accordance with their ability to enforce workers' effort, while considering technological and social conditions, it is not surprising that we do not observe any significant inefficiency in share tenancy. Then the prevalence of owner-cultivation in Asia can be explained mainly by the relatively equal distribution of land ownership rather than by the inefficiency of tenants' cultivation.

6.4 Consequences of Tenancy Regulations

Therefore, significant inefficiency of share tenancy is expected to arise particularly when the scope of contract choice is institutionally restricted. Indeed, case-studies reporting significant Marshallian inefficiency are clustered mainly in some states of India and Bangladesh in which contract choice is limited by land-reform regulations. As explained previously, the land-to-the-tiller legislation in Asia generally applies only to tenanted land. One consequence of this legislation was to induce a shift from the tenancy to the permanent-labour contract. Another consequence was to induce a shift from fixed-rent to share tenancy because the latter can be disguised easily as a labour-employment contract, even though share tenancy is prohibited by law in India. Indeed, 'with sharecropping it is not too difficult for the landlord to claim that the product belongs to him and that he is making from that product a payment to the tenant in return for his labor' (Bliss and Stern 1982: 131). In order to prevent tenants from establishing a claim that they were the actual tenants on any particular plot of land, landlords prevented them from cultivating the same plot from year to year or even crop season to season (Thorner and Thorner 1962; Dantwala and Shah 1971; Breman 1974; Ladejinsky 1977; Subbarao 1987). Thus, the opportunity for tenants to enjoy secure long-term tenancy contracts evaporated with the result of 'effectively increasing the management and supervision costs of the landlord' (Cain 1981: 447–8). According to the National Sample Survey, the ratio of tenanted land under share contracts at the all-India level increased from 36.6 per cent in 1953/4 to 48.0 per cent in 1970/1, while the ratio of farm area under tenancy decreased from 20.3 per cent to 11.6 per cent.[6]

It is in such an institutional environment that significant Marshallian inefficiency was found by the well-known studies of Bell (1977) and Shaban (1987) through a comparison of output and inputs between owned and share-cropped land plots in the same farms. To the extent that the tenant's ability is controlled on the same farm, the Bell–Shaban studies are considered reliable. However, their studies are subject to a limitation of not comparing production efficiencies between share and fixed-rent tenancy. This limitation is presumably due to too few observations of fixed-rent tenancy in their data for meaningful statistical comparison, because fixed-rent tenancies were the more overt form of tenancy and hence rarely practised in their study sites because of its legal prohibition. The fact that neither fixed-rent nor long-term share tenancy was the available option in both Bell's study site (Bihar) and Shaban's (semi-arid states in the Decan Plateau) has been confirmed by Ladejinsky (1977), Cain (1981), Jodha (1984), and Walker and Ryan (1990). The significant inefficiency of share tenancy is likely to arise from the landlord's choice of short-term lease to guard against the loss of land through long-term lease. Thus their findings represent strong evidence for the inefficiency of share tenancy to emerge under institutional constraints on tenancy choice rather than evidence for the inefficiency of share tenancy generally. In fact, similar to Bell's study, Bliss and Stern (1982: 133–8) compared yield between owned and sharecropped plots of the same farms and found no significant inefficiency of share tenancy in an area which was not subject to the significant impacts of land reform.

For Bangladesh, a number of case-studies exist reporting significantly lower rice yields per hectare of land under share tenancy than owner-farming (Jabbar 1977; Mandal 1980; Talukdar 1980; Shahid and Herdt 1982; Bhuiyan 1987; Bhuiyan and Nandan 1987).[7] Similar to Bell and Shaban's for India, most of these studies compared share tenancy with owner-farming because fixed-rent tenancies were rarely observed in Bangladesh (Hossain 1978). The land-reform laws of Bangladesh are similar to those of India except that in Bangladesh fixed-rent tenancy is prohibited but share tenancy is allowed as a form of labour contract, whereas the reverse is the case in India (Abdullah 1976). Yet the impact of these laws was the same in the sense that fixed-rent tenancy was precluded from the scope of contract choice, as evident from the data in Table 1.2. Even absentee landlords whose cost of monitoring tenants' work-effort is high were forced to adopt share tenancy. Significant Marshallian inefficiency is naturally expected in such an institutional environment.

Yet compared with the case of many states in India in which share tenancy has been prohibited, official approval of the practice of share tenancy would have improved efficiency in resource allocations in Bangladesh, especially when more intensive farming systems were introduced requiring greater care and judgement in farm-workers' operations. Indeed, a recent study on rice-farming across villages of different production environments found a clear tendency to substitute share tenancy for permanent-labour contracts as high-yielding modern varieties were adopted (Hossain and Akash 1990). It seems reasonable to hypothesize that production efficiency with new rice technology would have been constrained by lower work-incentives of the farm-workers if such a substitution were not allowed to take place.

6.5 Spectrum of Contract Choice

Those findings indicate the danger of drawing implications from the results of partial comparisons among observed contract options without due consideration of institutionally precluded alternatives. For example, Bell (1977) and Shaban (1987) may well be correct in finding relative inefficiency of share tenancy compared with owner-farming. However, their findings should not be taken as evidence of general inefficiency associated with share tenancy without considering possible resource allocation under permanent-labour contracts as well as legal constraints on the use of fixed-rent tenancy. Otherwise, they will be used as a justification for the prohibition of share tenancy in favour of owner-farming stipulated in land-reform laws in India and many other countries in Asia rather than relaxation of present tenancy regulations suppressing the fixed-rent tenancies.[8] Although the relative efficiency of share tenancy and permanent-labour contracts has not been examined so far, except in our case-study in the Philippines (Chapter 9), the very fact that many landlords in India have dared to practise illegal share tenancy under the guise of a labour contract is considered prima-facie evidence of the lower efficiency of fixed-wage permanent-labour contracts compared to share-tenancy contracts. If so, possible reinforcement of present tenancy regulations will be counter-productive to the efficient use of resources in the economy.

However, the available empirical evidence should not be interpreted as implying that the fixed-wage labour contract is universally less efficient than the share-tenancy contract. If some landlords have the ability to monitor tenants' work-effort, they can achieve the same degree of efficiency with share tenancy as with fixed-rent tenancy. Similarly, there may be exceptionally capable landlords who can achieve the same efficiency with the labour of fixed-wage employees. According to Spillman's (1919) theory of the 'agricultural ladder', land-tenure choices have a life-cycle element: a young boy seeking a farming career begins employment as a farm-helper or permanent labourer and learns farming techniques from an experienced master farmer; next he becomes a share tenant and receives advice from his well-informed landlord; as his farming knowledge and financial capacities increase, he ascends to the status of fixed-rent leasehold tenant and, if successful, then to owner-farmer; as he becomes older, he retreats from farm-work first by employing farm-helpers and later by renting his land out to tenants. In this framework, Reid (1973, 1975, 1976a, 1976b, 1979a, 1979b) interprets the prevalence of share-cropping in the *post bellum* US South as the efficient means of combining the labour of poor and inexperienced workers with the farming knowledge and the financial capacity of landlords. Although Reid does not formalize his argument, the incentive structure envisaged by him seems to be captured by Eswaran-Kotwal's model of share tenancy reviewed in Section 4.3. If the monitoring cost of the worker's effort is sufficiently small, an arrangement whereby a young farm-helper acquires human capital under a well-experienced master farmer can be a fully efficient contractual system.[9]

The problem in South Asia is that access to this ladder is made difficult by the caste system and land-reform legislation. Because the land rental market is institutionally repressed, a farmer with large endowments of land, animals, and machines relative to family labour needs to employ permanent labourers in order to exploit

Global Survey of Empirical Evidence 105

the limited scale economies (Section 1.2). Since his land endowment is largely hereditary, it is usually not proportional to his farming experience and skill. Where his managerial ability to supervise hired labourers is unmatched with other factor endowments, inefficiency is bound to arise. This difficulty is exacerbated by the hired labourers' poor incentives; they have little likelihood of ascending the agricultural ladder by following their master's guidance.

In South-East Asia, share tenancy is commonly used as a step for farmers' sons (or sons-in-law) to inherit the farms of their parents or relatives, but permanent-labour contracts are rarely found (Hirashima 1977; Kuchiba and Banzon 1979; Kuchiba *et al.* 1979; Fujimoto 1983). More common is the case in which a newly married couple from farmer families live in a shanty near their parents' house and live as casual labourers working on their parents' and other neighbours' farms and are paid according to various time-rate, piece-rate, and team-work contracts for different tasks (Roumasset and Uy 1980; Hayami and Kikuchi 1982). This seems to indicate that, given strong family and community ties, fixed-wage casual labour can be managed efficiently.

Thus the available evidence is consistent with a general hypothesis that rural people in developing economies make efficient choices from a wide spectrum of contracts, ranging from casual-labour employment to long-term fixed-rent tenancy. This is consistent with T. W. Schultz's (1964) view that traditional agriculture allocates resources efficiently. However, as Basu (1983, 1984) argues, contracts in less developed agrarian economies tend to be personalized among relatives and friends and hence contractual opportunities tend to be geographically isolated. This consideration suggests a possibility that allocative efficiency in one small location does not assure overall efficiency across regions or in the economy as a whole. This hypothesis is yet to be examined empirically.

Notes to Chapter 6

1. For example, Bliss and Stern (1982) found only a few permanent labourers in a village in Uttar Pradesh where land reform was not practically implemented.

2. Examining the reliability of the National Sample Survey data, Narain and Joshi (1969) conclude that the incidence of tenancy declined as a result of the implementation of the land-reform legislation, even though the underreporting of tenancy contracts increased significantly with the aim of evading reform programmes.
3. It must be noted that the yield data taken from past studies are average yields of the sample rather than the original sample data, so that the distributions shown in Tables 6.1 and 6.2 are more in the nature of sampling distributions than the distribution of the sample. Also note that in order to control the effects of land quality and farm size on the yield differential, disaggregated yield data classified by the irrigation condition and the size of cultivation were used as far as possible.
4. Besides the literature cited in Table 6.1, Vyas (1970) as well as Chakravarty and Rudra (1973) reported no significant inefficiency under share tenancy though their data were not published.
5. Care must be taken in interpreting the comparison of labour inputs because they are measured in terms of working days. As was pointed out in Ch. 2, it is not the time the tenant spends but the 'effort' he expends (which, of course, is difficult to measure empirically) that is directly relevant to the tenancy model.
6. The National Sample Survey reports a larger incidence of tenancy than the National Census data shown in Table 1.2.
7. The mean or the rate of difference in physical yield between share tenancy and owner-farming based on the data provided by Bangladesh studies is -11%, which is significantly different from zero. If we remove Bangladesh samples from the physical yield comparison in Table 6.1 and Fig. 6.1, the distribution becomes more symmetrical. However, the major qualitative properties remain unchanged.
8. For example, Bell's research result provides the basis of the argument for land-reform programmes proposed by Prosterman and Riedinger (1987).
9. The theory of the agricultural ladder suggests the importance of human capital formation as a determinant of one's lifetime earnings profile in agriculture. Rosenzweig and Wolpin (1985) provide empirical evidence consistent with this theory; given the heterogeneity of land plots, young farmers who embark on farming careers acquire valuable farming knowledge specific to plots of land from their fathers and from their own experience.

7

Contract Choice and Enforcement in an Agrarian Community
The Case of Upland Farming in Indonesia[1]

THE theory of contract choice based on our basic model (Chapters 2 and 3) predicts that, where stipulated contract terms are difficult for landowning principals to enforce and landless farm-workers are risk-neutral, the fixed-rent tenancy contract will be chosen; when workers are risk-averse, an increase in the worker's output-sharing rate enhances his work-incentives only at the cost of increasing his income risk and hence given such a trade-off the share-tenancy contract will be chosen in equilibrium. To the extent that the worker's incentive is smaller under share tenancy, his work-effort is expected to be less than optimum. Yet the comparisons in the levels of input and output per hectare among different land-tenure forms in Chapter 6 did not as a rule find significant inefficiency associated with share tenancy where both share and the fixed-rent tenancy contracts were available options. It is hypothesized that inefficiency is usually not so significant because landlords base their contract choice on the rational calculation of their abilities to monitor their tenants' work-effort, consonant with tenants' rational choice based on their risk-bearing as well as managerial abilities. The fact that significant inefficiency is found to be associated with share tenancy mainly where the option of fixed-rent tenancy is institutionally precluded is consistent with the general hypothesis that rural people in developing economies tend to make a socially efficient choice from a wide spectrum of agrarian contracts if they have freedom to choose.

In order to provide more concrete evidence in support of this hypothesis, we present in this chapter a case-study on the choice and enforcement of land-tenancy contracts in a village in Indonesia. Land-tenure contracts in developing economies are highly complex and elusive as they are based mostly on tacit agreements rather than formal contracts. Contractual terms are mutually understood often without formal oral statements, not to speak of written contracts. Moreover, landlords are usually suspicious and do not respond honestly to investigations by outsiders on land-tenure issues. The difficulty in data collection is increased by a large number of small scattered plots of a landowner being cultivated usually by himself and several tenants with different cropping systems under different tenure arrangements. For this reason, we limited our analysis to one small location while sacrificing national or regional representativeness. Our strategy was to conduct a complete survey of all landlords, owner-cultivators, and tenants in one village (*kampung*), collecting data on all individual plots.

In this village, virtually no artificial constraint is effective on land-tenure arrangements. In Indonesia, the Basic Agrarian Law of 1960 forbids absentee landownership and sets maximum limits for private holdings according to population density. However, these limits were set so high that the impact of the land-redistribution programme has been confined to plantations (mostly foreign-owned), and the peasant sector, including this study village, has been unaffected. Tenancy regulations based on both this law and the Sharecropping Law have been a dead letter, especially since the bloody suppression of communists in 1965 (Utrecht 1969; Hayami and Kikuchi 1982: 152–3). Moreover, in this village there is no social constraint like the caste system in India to constrain the land-tenure choice. Therefore, the case-study of this Indonesian village will illuminate how the choice is made from a wide spectrum of agrarian contracts where the choice is unrestricted.

This village study will also provide insights into contract choice and enforcement in the production environment in which complex intercropping systems including cash crops are practised. The crop yield and input comparisons in the past between share tenancy and other land-tenure forms (reviewed in Chapter 6) pertain mostly to cases of single cropping of major food staples such as rice and wheat for which the monitoring of tenants' inputs is relatively easy either directly or indirectly through the observations of yield outcomes. Monitoring tends to become more difficult for more complex cropping systems or crop–livestock combinations. Moreover, tenants' inputs are more difficult to monitor for some commercial crops such as tobacco, of which both the quantity and the quality of output are very sensitive to the input of the managerial and entrepreneurial ability of tenants that is hard to measure. This difficulty is considered to underlie the lower incidence of share tenancy in cash than in staple food crops despite higher risks involved in the former (C. H. H. Rao 1971). It is also considered to underlie the phenomenon that high-value cash crops are grown more commonly in owned than share-cropped plots in the same farms (Bell 1977). Therefore, even though significant Marshallian inefficiency has not been found in relatively simple production systems of major food staples, it might be found in more complicated cropping systems including cash crops, for which little investigation has been carried out. This study intends to shed light on this unresolved issue.

Above all, this case-study attempts to illustrate the effective mechanism of monitoring and enforcing tenants' work-effort in an agrarian community where social interactions are intense and transactions are expected to continue over long periods. For both environmental and historical reasons, villages in Indonesia, especially in old-settled Java in which this study village is located, are known to be characterized by intensive social interactions and strong community ties (Furnival 1944; Geertz 1970, Hayami and Kikuchi 1982). In such a social environment, reputation is likely to play a major role in contract enforcement in the context of long-term contracts, as discussed in Chapter 4.

7.1 Characteristics of the Study Village

The village (*kampung*) chosen for this study is located in the Garut District in West Java. In this village, a research project on an integrated analysis of farm production and household economy was conducted for two full years from January 1985 to December 1986 (Morooka *et al.* 1989). Through this project, we were able to establish mutual trust with villagers and to collect reliable data on farm outputs and inputs. The data pertaining to land tenure were collected by a special survey conducted in July 1985.

This village is a typical upland village on a hilly plateau in which various crops are grown in terraces under rainfed conditions. Most villagers are Sundanese and pious Moslems. An unpaved road of about one kilometre connects the central part of this village to a national highway that runs from the Garut City to Bandung City, the capital of West Java (see Fig. 7.1). The central market (bazaar) of Garut is 8 kilometres away and can be reached by pony wagon or minibus in about 20 minutes.

Landholdings

Farming is the main occupation in this village, while many farmers and their family members also engage in other occupations such as petty trades, transport, and construction work. As shown in Table 7.1, about one-quarter of households has no farmland and another

TABLE 7.1. *Size distribution of land ownership and operational holdings by villagers in the study village, Garut, Indonesia, 1986*

	Ownership holding		Operational holding	
	No.	%	No.	%
0	35	24	24	17
0.01–0.2	36	25	31	21
0.21–0.6	40	28	56	39
0.61–1.0	24	17	22	15
1.01–2.0	9	6	11	8
2.01	1	—	1	—
TOTAL	145	100	145	100
Total area (ha.)	54.2		61.5	
Average of all households (ha.)[a]	0.37		0.42	
Average of farm households (ha.)	0.45		0.51	

[a] Excludes 4 householders who did not reply to our interview.

one-quarter owns less than 0.2 ha. The average area of farmland owned by all households is less than 0.4 ha. and the average for farm households is only about 0.5 ha. Landless villagers make a living either as tenants or as agricultural labourers with their income supplemented by casual non-farm work. Operational farm sizes are also very small with an average of 0.5 ha., and the farmers cultivating more than 1 ha. are only 8 per cent. Farming is of a typical peasant mode based mainly on family labour with the aid of

The Case of Upland Farming in Indonesia 111

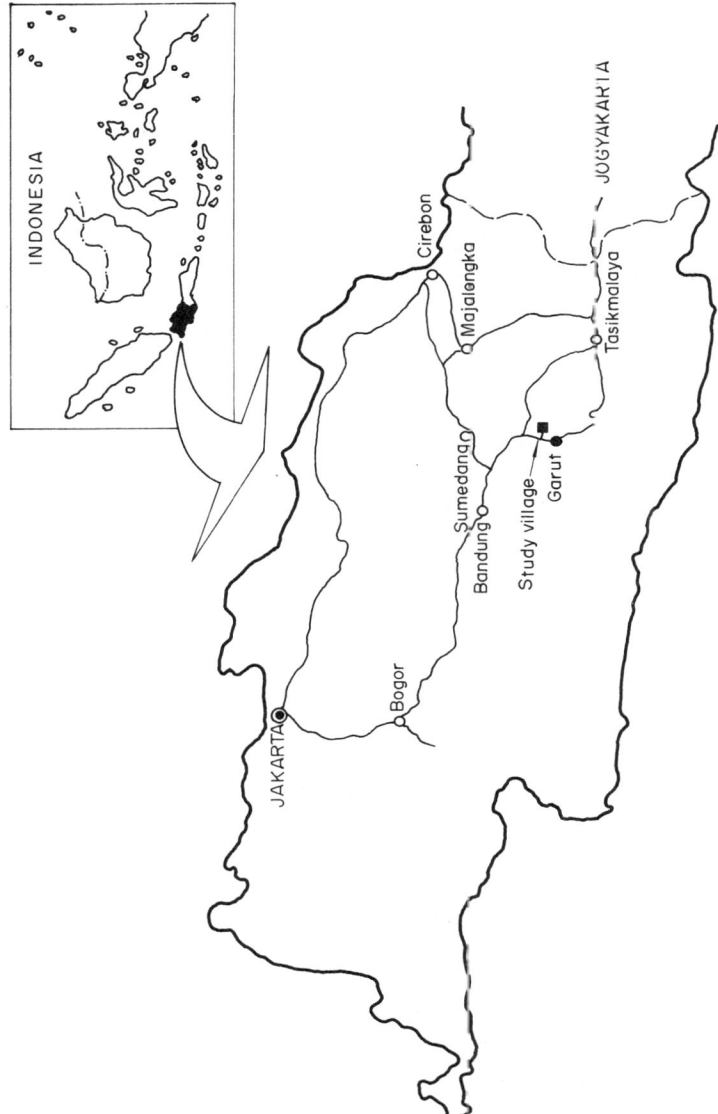

FIG. 7.1 The Garut District in West Java, Indonesia

hired or exchange labour at busy seasons such as harvesting. Hired labour is employed exclusively under casual-labour contracts by specific task, with remuneration based on time- (daily or hourly) rates or piece-rates. No incidence of permanent labour employed by year or by crop season for a wide variety of tasks is observed.

Farming systems

This village is surrounded by upland terraces in rainfed conditions, where various crops are interplanted. A typical cropping system shown in Fig. 7.2 mixes soybean and corn together for the first crop (September–December), followed by soybean and tobacco

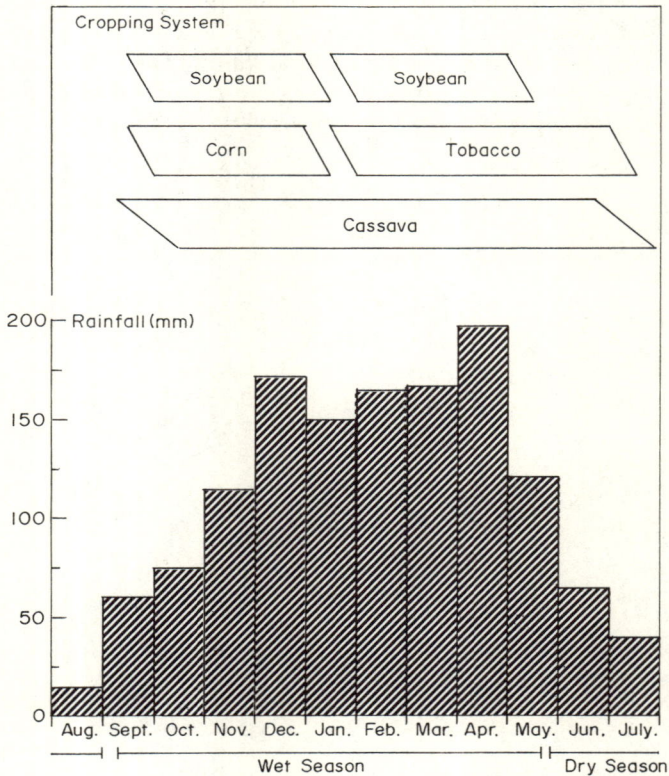

Fig. 7.2 Average monthly rainfall and a typical cropping system in the study area

for the second crop (January–June). In addition, cassava is often planted year-round on the edge of the fields. There are many other variations, e.g. upland rice is planted instead of soybean or corn is planted instead of tobacco for the second crop. The intercropping of such *palawija* crops is common to the upland areas of not only Java but also monsoon Asia in general. Land preparation is done manually with hand hoe and no draught animal is used for upland farming in this area.

7.2 Land-Tenure Systems

As is common in Java, owner-cultivation is the dominant form of land tenure in this village, occupying 73 per cent of 256 land plots and 72 per cent of 57 ha. of agricultural land (Table 7.2). Out of 121 farmers, 84 per cent are owner-operators or owner/tenant-operators and pure tenant-operators are only 16 per cent.

TABLE 7.2. *Distribution of farm land, farm-operators, and landlords by tenure status in the study village, Garut, Indonesia, 1986*

	Owner-cultivation	Tenant-Cultivation				Total
		Fixed	Pawn	Share	Total	
No. of plots	188	18	19	31	68	256
(%)	(73)	(7)	(8)	(12)	(27)	(100)
Land area (ha.)	40.9	6.4	2.9	6.8	16.1	57.0
(%)	(72)	(11)	(5)	(12)	(28)	(100)
No. of farms-operators	102[a]	16	16	23	57[b]	121[c]
(%)	(84)	(13)	(13)	(19)	(47)	(100)
No. of landlords	—	16	18	27	16	—
(%)	—	(26)	(30)	(44)	(100)	—

[a] Includes owner-tenant operators.
[b] The numbers of tenants in different tenure classes do not add up to the total because some tenants cultivate different plots under different tenancy contracts.
[c] The numbers of owner- and tenant-operators do not add up to the total because some operators cultivate both owned and rent-in land plots.

Forms of tenancy contract

There are three forms of land-tenancy contracts. The most common is share-cropping, occupying more than 40 per cent of tenanted land plots as well as area. Two types of share tenancy are practised. In an arrangement called *maro*, outputs and current inputs such as seeds, fertilizers, and chemicals are typically shared equally between landlords and tenants, while all other inputs are borne by tenants. In another arrangement called *mertilu*, tenants are entitled to receive two-thirds of output while they shoulder all the costs. (For the distribution of output and cost-sharing arrangements, see Table 7.3.)

Another common form of tenancy is the fixed-rent contract (*sewa*), by which a fixed amount of cash is paid in advance at the beginning of a crop year (around August). This rent-payment schedule in upland agriculture is very different from the common practice in lowland rice-farming in Java, in which a fixed rent is paid usually in kind at the time of harvest. The practice of advance rent payment in this area rules out the possibility that share tenancy is preferred to fixed-rent tenancy by landlords as a device to prevent tenants' defaulting on rent payments (Allen 1985).

A unique form of tenancy found in this area is the pawn contract (*gadai*), by which a tenant establishes the right to continue cultivating a land plot for a certain money deposit to its owner until the borrower pays back the loan. An implicit interest to be generated from the deposit is considered the land rent. This contract serves as an important source of informal credit in an area in which explicit charge of interest is prohibited by the Muslim tradition.

Written contracts were exchanged for all the cases of land-pawning that we observed, while no written documents were exchanged in the case of both share and fixed-rent tenancy. This difference seems to reflect the unique nature of pawning as a credit institution as well as a land-tenure institution. Besides the pawn contract, no other tenancy contract was reported to be interlinked with credit contract. Furthermore, no contract stipulated tenants' obligations to provide unpaid labour services to landlords. All the contracts can be terminated at the end of each crop year, though they commonly continue to be renewed for some years.

TABLE 7.3. *Distribution of cost- and output-sharing arrangements for soybean, corn, and tobacco by plot under share tenancy (maro and mertilu) in the study village, Garut, Indonesia, 1986 (no. of plots)[a]*

Tenants' cost share (%)	Tenants' output share		Total
	maro 1/2	mertilu 2/3	
Seeds:			
100	5	17	22
50	2	0	2
0	5	0	5
TOTAL	12	17	29[b]
(%)	(41)	(59)	(100)
Fertilizers:			
Urea			
100	2	16	18
50	5	0	5
0	5	1	6
TOTAL	12	17	29
(%)	(41)	(59)	(100)
Manure			
100	3	14	17
50	2	0	2
0	7	3	10
TOTAL	12	17	29
(%)	(41)	(59)	(100)

[a] In case of cassava, seedlings were shouldered by tenants. Percentages of arrangements by which tenants received total, two-thirds, one-half of cassava output were, respectively, 15%, 63%, and 22%.
[b] Exclude 2 plots due to lack of data.

Characteristics of landlords and tenants

Various factors underlying the choice of land-tenure arrangements are suggested from the average characteristics of landlords and tenants summarized in Table 7.4. On average, landlords owned a larger area of land (1 ha.) than tenants (0.2 ha.). The landlords also tend to be older and the percentage of female landlords to be

TABLE 7.4. *Characteristics of landlords and tenants by tenancy contract in the study village, Garut, Indonesia, 1986*

	Fixed	Pawn	Share	Average
Landlord characteristics:				
Age of family head (yrs.)	47	47	54	49
Ratio of female family head (%)	31	11	22	21
Land (ha.):				
Owned	1.4	0.6	1.1	1.0
Rent-out	1.3	0.2	0.5	0.6
Self-cultivated[a]	0.2	0.4	0.6	0.4
Tenant characteristics:				
Age of family head (yrs.)	39	43	40	41
Ratio of female family head (%)	6	0	13	7
Land (ha.):				
Owned	0.3	0.3	0.2	0.2
Rent-in	0.4	0.2	0.3	0.3
Total cultivated[b]	0.7	0.5	0.5	0.5

[a] (Owned) − (Rent-out)
[b] (Owned) + (Rent-in)

much higher than the tenants. Such observations seem consistent with the hypothesis that tenancy contracts were made for the purpose of equilibrating marginal labour productivities by equalizing effective labour inputs per hectare between landlords and tenants. No incidence of permanent-labour employment seems to indicate that landlords prefer land tenancy to permanent-labour contracts presumably because of the very high cost of supervising hired labourers for dispersed and diverse farm operations even in a small community with strong social interactions (Chapters 1 and 4).

Thus considered, the contract choice may have a life-cycle dimension. However, the hypothesis advanced by Reid (1976b) that share tenancy tends to be preferred by young inexperienced farmers as an earlier step to climb up the 'agricultural ladder' via fixed-rent tenancy does not seem to apply to this village because the average age of share tenants was not lower than that of fixed-rent tenants (Section 6.5).

A more critical consideration in the choice of tenancy contracts seems to be the cost of enforcing contractual terms. Indeed, the

The Case of Upland Farming in Indonesia

landlords who adopted fixed-rent contracts had a higher female ratio and rented out a larger area than the landlords who adopted share contracts; this seems to reflect both the much smaller capacity of female (widow) and scale diseconomies in the supervision of tenants' farm operations.

Land areas both owned and rented out by the landlords who adopted the pawn contract were very small, despite the contract-enforcement cost being considered to be no different from that of the fixed-rent contract. This fact seems to reflect a more frequent use of land-pawning for credit purposes by smaller (and poorer) landowners. It appears that the choice of the pawn contract was dominated by considerations of need for credit.

7.3 Relative Efficiencies and Enforcement Costs

Observations in the previous section suggest that it is not costless in this village community to enforce tenants' work-effort. If so, significant Marshallian inefficiency may be associated with share tenancy. However, if share contracts were adopted only by landlords with ability to enforce tenants' work-effort at a low cost and those not possessing such ability (e.g. widowed female landlords who have little knowledge on farming practices) chose fixed-rent tenancy, no significant inefficiency would be observed for share tenancy relative to fixed-rent tenancy and owner-cultivation.

Comparisons in outputs and inputs per hectare

In order to resolve this question, average output and input per hectare of land under share tenancy are compared with those under fixed-rent tenancy and owner-cultivation (Table 7.5). In making comparisons, observations for fixed-rent and pawn contracts are grouped together because these contracts are considered similar in terms of incentives for agents (tenants) to apply their inputs.

The output comparison is made in terms of total value of output per hectare per year, aggregating individual crop yields at market prices at the farm-gate. In addition, a comparison is made in terms of gross value added after subtracting current input value from total output value. Comparisons by individual crops are not

TABLE 7.5. *Average values of gross output, gross value added, current input (Rp. 1,000/ha.), and labour input (Days/ha.) by tenure class in the study village, Garut, Indonesia, 1986*

	Gross output	Value added	Current input		Working days
			Fert. and chemical	Total	
Owner (153, 43)[a]	860	658	133	202	232
	(627)[b]	(577)	(117)	(133)	(83)
Fixed (34, 7)	890	705	122	185	262
	(836)	(768)	(109)	(117)	(96)
Share (27, 11)	831	624	126	207	208
	(673)	(614)	(104)	(126)	(63)
t-statistics[c]					
Owner vs. fixed	0.195	0.337	0.482	0.703	0.860
Owner vs. share	0.220	0.279	0.301	0.174	0.890
Fixed vs. share	0.296	0.446	0.121	0.711	1.435

[a] Sample size: the first figure refers to the number of observations by plot for the tests for output, value added, and current input and the second figure refers to the number of observations by farm for the tests for working days. Working days were evaluated for farmers who used similar types of cropping system under the same tenure status throughout the year.
[b] Standard deviations are shown in parentheses.
[c] All the differences are not significantly different from zero at the 10% level.

attempted because the yields of individual crops planted in intercropping systems are in a trade-off relation and hence they are not relevant indicators of production efficiency. Comparisons of inputs are made in terms of the values of current inputs, both the aggregate of fertilizers and chemicals and the total including seeds and seedlings, as well as of labour measured in workdays per hectare per year.

Student t-statistics calculated for testing the null hypothesis on the equality in output and input levels between share tenancy and owner-cultivation and between share and fixed-rent (and pawn) tenancy show that in no case can the null hypothesis be rejected even at the 10 per cent significance level. Such results represent strong evidence in support of the hypothesis that equal allocative efficiency was achieved by share tenancy as compared with owner-cultivation and fixed-rent tenancy.

Distribution of cropping systems by contract

Equal efficiency among forms of land-tenure was not associated with any systematic difference in the choice of crops and cropping systems. It may be expected that single cropping is preferred to more complicated intercropping under share tenancy because there is more scope in the latter for changing the input of tenants' managerial and entrepreneurial ability which is difficult for the landlord to monitor. For the same reason, high-value commercial crops such as tobacco may be preferred less than staple food crops under share tenancy, as argued by C. H. H. Rao (1971). Indeed, tobacco cultivation is known to be characterized by very large variations in revenue due to different natural conditions including soil, rainfall, and pest infestation as well as in growers' technical skill and management ability.

However, comparisons in Table 7.6 show that both the practice of intercropping and the planting of tobacco were no less common

TABLE 7.6. *Distribution of plots among different cropping systems for different land-tenure classes in the study village, Garut, Indonesia, 1986 (%)*

	Owner-cultivation ($n=188$)[a]	Tenant-cultivation		Total ($n=256$)
		Fixed/pawn ($n=37$)	Share ($n=31$)	
Case I:				
Intercropping	85	92	94	86
Single cropping	15	8	6	14
TOTAL	100	100	100	100
Case II:				
Planting tobacco	49	49	55	48
Not planting tobacco	51	51	46	52
TOTAL	100	100	100	100

[a] The total number of plots in each tenure class.

under share tenancy than under owner-cultivation and fixed-rent tenancy. The null hypothesis that the choice of cropping systems was not contingent on the choice of class of land-tenure is accepted at the 5 per cent level of significance on the basis of chi-square tests, with the calculated chi-squares being 2.475 for the comparison between intercropping and single-cropping systems (case I) and 1.076 for the comparison between systems planted with and without tobacco (case II).

The results of Tables 7.5 and 7.6 are consistent with the hypothesis that landlords who chose share tenancy were able to enforce contractual terms even with complicated cropping systems, including high-value commercial crops, and that those who did not possess the ability to do so chose fixed-rent tenancy.

Comparisons in land rent

A major question is how high the cost of contract enforcement would be. This cost, plus a premium for risk to be shouldered by landlords, should present a margin of share rent over fixed rent. Therefore, the magnitude of the enforcement cost can be inferred from a comparison between share and fixed rent.

First, the output to be shared between landlords and tenants under share tenancy is subject to fluctuations in yield due to variable natural conditions. This is not a serious problem for comparisons in output because it can be reasonably assumed that plots under share and fixed-rent contracts were on average subject to the same fluctuations. However, since fixed rent paid in advance must have been determined on the basis of normal crop yields, comparable share rent must also be calculated from these yields. Unfortunately, we do not have sufficient time-series to estimate normal yields. We gathered that soybean yields for the first crop of the survey year were abnormally low due to serious damage caused by bean fly (*Ophiomia phaseoli*). In order to adjust for this, we substituted the average soybean yield in the second crop for that of the first crop, although this adjustment may still underestimate the normal yield because the first-crop soybean yield is usually higher than that of the second crop.

Secondly, to be comparable with fixed-rent paid in advance, the current market value of crops produced and inputs purchased over a year under share tenancy must be discounted back to a present value at the beginning of a crop year. The problem is what discount rate should be used. In the absence of an explicit interest rate in the Muslim community, a relevant rate was estimated from rents under

The Case of Upland Farming in Indonesia 121

the fixed-rent and the pawn contracts. It was found that the average deposit required to use a hectare of land under the pawn contract was 1,575,000 rupiahs, implying that rent under this contract was this sum multiplied by an interest rate (r). On the other hand, the average rent under fixed-rent tenancy was found to be 224,000 rupiahs per hectare. Since the fixed rent was paid in advance, an equilibrium between the fixed rent and the pawn rent is represented by:

$$224 (1 + r) = 1,575 r.$$

By solving the above equation, the interest rate (r) is estimated to by 17 per cent per year or 1.3 per cent per month. This rate is considered an opportunity interest rate that landlords with the share contract could have earned if they had opted for the fixed-rent or pawn contract.

Using this interest rate, the current market values of outputs and inputs under share contracts were discounted back to the present value at the beginning of the crop year on the assumption that the first crop, the second crop, and cassava were harvested four, ten, and twelve months respectively after planting and that the current inputs for the first crop and cassava were purchased at the beginning of the crop year and those of the second crop six months later.

An average share rent thus estimated is compared with an average fixed rent in Table 7.7. It is found that the current value of share rent under the *maro* arrangement was almost 60 per cent higher than fixed rent but the discounted value at the beginning of a year was about 40 per cent higher. In the *mertilu* case, in which landlords do not share the cost, the discounted share rent was only about 20 per cent higher than the fixed-rent. The higher rent under the *maro* than under the *mertilu* arrangement is considered to reflect the higher risk that landlords have to bear under the cost-sharing contract. It must be noted that our calculation tends to underestimate the true margin of share rent over fixed rent because our adjustment for soybean damage to the first crop is likely to be insufficient.

With our present knowledge, it is not possible to separate the estimated difference in rents between the cost of share-contract enforcement and a premium for risk-sharing. However, considering the relatively high risk involved in uni-rrigated upland farming, a premium for share tenancy that landlords would request and that

TABLE 7.7. *Comparison of estimated share rent with fixed rent (Rp 1,000/ha.)*

	1st crop	2nd crop	Cassava	Total maro	mertilu
Output:					
Current value (1)	221	674	17	912	912
Discounted (2)	210	592	15	817	817
Current input cost to be shared:					
Current value (3)	37	169	1	207	0
Discounted (4)	37	156	1	194	0
Income to be shared:					
Current value (1)−(3)	—	—	—	705	912
Discounted value (2)−(4)	—	—	—	623	817
Share rent (5):					
Current value	—	—	—	353	304
	—	—	—	(158)[a]	(136)
Discounted value	—	—	—	312	272
	—	—	—	(139)	(121)
Fixed rent				224	
				(100)	

Notes: (1) Market value at the time of harvest while substituting 2nd crop soybean yield for 1st crop yield (see text).
(2) Present value at the beginning of crop year by discounting current market values with an interest rate of 1.3% per month, assuming 1st crop, 2nd crop, and cassava harvested 4, 10, and 12 months later respectively.
(3) Market value at the time of purchase.
(4) Present value obtained by the same procedure as for (2), assuming inputs for 1st crop and cassava purchased at the beginning of crop season and those for 2nd crop purchased 6 months later.
(5) Share rents for *maro* and *mertilu* are derived from value added multiplied by 1/2 and 1/3 respectively.

[a] Percentages to fixed rent are shown in parentheses.

tenants would agree to pay would not be small. It does not seem unreasonable to assume that the difference between the *maro* rent and the *mertilu* rent is explained mainly by risk premium; this means that the cost of share-contract enforcement is relatively modest.

It must be pointed out that the interest rate estimated from fixed-rent and pawn deposit was based on safe collateral. The effective interest rate for landless people lacking good collateral would have been substantially higher. For example, in some lowland rice villages in West Java, where interest is explicitly charged rather exceptionally, the interest rate for one rice crop season amounted to 50 per cent, implying that the annual rate is higher than 100 per cent (Hayami and Kikuchi 1982: 200; Fujimoto 1986: 91). Also, in an informal credit arrangement called *kopia* (*Koperasi Simpan Pinjam*) commonly practised in Java, a person who borrows 1,000 rupiahs has to pay back to the lender 40 rupiahs every day for one month, implying that the interest rate is higher than 20 per cent per month. It does not seem unreasonable to assume that, in our study site, too, the market rate of interest for a non-collateral loan to which landless tenants have access can be as high as 100 per cent (6 per cent per month) or higher.

The present values of share rent discounted at an interest rate of 100 per cent instead of 17 per cent per year are found to be 212,000 and 224,000 rupiahs per hectare for the *maro* and the *mertilu* cases respectively, which are almost exactly the same as the fixed rent of 224,000 rupiahs. These results imply that, while effective rent under share tenancy is higher than that of fixed-rent tenancy for landlords who have access to collateral, the share rent may not be higher or can even be lower than the fixed rent for landless tenants for whom only high-interest non-collateral loans are available. Such different opportunity costs of credit faced by landowners and the landless are thus considered to be one of the major factors underlying the choice of share tenancy, especially in an imperfect credit market where an explicit interest charge is prohibited.

7.4 Community Mechanism of Contract Enforcement

What mechanism would exist in enforcing the terms of share contracts in this village community? The direct supervision by landlords on tenants' work is an obvious possibility and, in fact, was practised to some extent. A majority of landlords adopting the share contract reported that they did pay visits to the plots they had rented out to monitor tenants' activities, while none paid such visits under the fixed-rent and the pawn contracts (Table 7.8).

TABLE 7.8. *Frequency of landlords' visits to rented-out plots (no. of landlords)*

Frequency	Fixed	Pawn	Share
0	16	18	12 (44)[a]
1/crop season	0	0	7 (26)
1/month	0	0	0 (0)
2 to 3/month	0	0	4 (15)
4 to 5/month	0	0	4 (15)
TOTAL	16	18	27 (100)

[a] Percentages are shown in parentheses.

Judging from the frequency of visits, landlords' direct supervision does not appear to be quite as intensive in this village as compared with others (Nabi 1986). However, the purposive visits reported in Table 7.8 would probably underestimate landlords' monitoring efforts. Informal discussions revealed that they tried to collect information by various means such as casual observations on rented-out plots on the way to their own cultivated fields or on other errands and reports from friends and neighbours who happened to pass by the plots. In a small community characterized by intense personal interactions, tenants' activities can be traced relatively easily if landlords are ready to listen to gossip. How accurately they can estimate tenants' inputs from such indirect evidence would depend on the landlords' experience and knowledge of farming. For landlords who have the ability to make accurate estimates, the proper enforcement of tenants' work-effort under share contracts could be achieved with little use of the resources of high opportunity cost.

Another way of solving the problem of contract enforcement would be to develop 'the relations of trust and confidence between principal and agent . . . so that the agent will not cheat even if it may be "rational economic behaviour" to do so' (Arrow 1968:

The Case of Upland Farming in Indonesia

538). In this connection, personal ties among family members, relatives, and friends are valuable for efficient contract enforcement (Ben-Porath 1980; Pollak 1985). Such personal ties will not only make tenants morally obliged to conform to the terms of contract but also increase the penalty for cheating and shirking since discovery of dishonest behaviour will not only deprive them of a mutual-help relationship with their present landlord but also reduce their chance of developing such a relationship with other landlords in the same community. Indeed, a much higher percentage of share contracts was found among family members, relatives, and friends than of fixed-rent and pawn contracts (Table 7.9).

TABLE 7.9. *Personal relations between landlords and tenants (no. of plots)*

	Fixed	Pawn	Share
Family/relative	10	9	24
	(56)[a]	(48)	(77)
Friend/neighbour	3	1	3
	(17)	(5)	(10)
Other	5	9	4
	(28)	(47)	(13)
TOTAL	18	19	31
	(100)	(100)	(100)

[a] Percentages are shown in parentheses.

Strong personal ties common to share tenancy were also reflected in a longer cumulative duration for the share contract than for the fixed-rent contract, although the duration of the pawn contract was usually very long as it continues until the loan is paid back (Table 7.10). The relatively long duration of the share contract may also reflect the fact that a tenant's inputs can be more clearly observed by his landlord because the influences of random factors such as weather will cancel out over many years. The longer the duration of the contract and the more intense the social interactions, the stronger the role of reputation in preventing parties to the contract from opportunistic behaviour, as discussed in Chapter 4.

It must be pointed out that all these efforts of landlords to enforce share contracts involve little opportunity cost. The fact that no

TABLE 7.10. *Cumulative duration of tenancy contracts (no. of plots)*

	Fixed	Pawn	Share
1 year	10	3	9
	(55)[a]	(16)	(20)
2 years	3	1	4
	(17)	(5)	(13)
3 to 5 years	3	4	7
	(17)	(21)	(23)
More than 5 years[b]	2	11	11
	(11)	(58)	(35)
TOTAL	18	19	31
	(100)	(100)	(100)

[a] Percentages are shown in parentheses.
[b] Includes cases of no answer due to lack of memory.

significant Marshallian inefficiency was observed (Table 7.5) reflects the ability of landlords to enforce the terms of share contracts at a low opportunity cost, while those who were not endowed with such ability would have chosen a fixed-rent tenancy.

7.5 Conclusion

In the upland area in West Java under investigation complex intercropping systems including high-value commercial crops were practised in risky unirrigated conditions. Despite the obvious difficulty for landlords to monitor tenants' inputs, the share-tenancy contract was commonly adopted side by side with the fixed-rent contract. No significant inefficiency in resource allocation was found to be associated with share tenancy as compared with owner-cultivation and fixed-rent tenancy. A comparison between share rent and fixed rent suggests that the cost of share-contract enforcement may not be so large. It was found that landlords paid substantial attention to contract enforcement. However, those efforts usually involved little input of the resources with high opportunity cost. These results reinforce our earlier conclusion drawn from the

The Case of Upland Farming in Indonesia

survey of past empirical literature in Chapter 6 that rural people in developing economies have sufficient ability to make efficient choice among alternative contracts with due consideration to both their resource endowments and the external conditions surrounding them.

Note to Chapter 7

1. This chapter draws heavily on Morooka and Hayami (1989).

8

Community and Market in Contract Choice
The Case of the Jeepney
in the Philippines[1]

THE previous chapter reports on contract choice and enforcement mechanism in a typical agrarian community characterized by strong personal ties and intensive social interactions, in which the cost to landowners of monitoring the work-effort of landless agents is relatively modest. Economic development is usually associated with a transformation from such community-based economies to more impersonal market-based economies. As people in a hitherto closed community are exposed to wide market opportunities and begin to engage in transactions with outsiders, personal ties will be weakened and social interactions within the community will be reduced thus weakening the community mechanism of contract enforcement. On the other hand, access to wide markets may reduce the risk associated with the transactions of both inputs and output. Thus the changes in risk and transaction costs resulting from the penetration of markets to agrarian communities will induce changes in the choice of contracts.

In order to illustrate this process, this chapter reports the results of a case-study on the choice of rental contracts of the 'jeepney'—an informal minibus—in the Philippines. The jeepney rental contracts are not 'agrarian' in that they do not combine land and labour for agricultural production but combine capital and labour for the production of transport services. However, analysis of the jeepney case is useful in illuminating the impact of market developments on contract choice in agrarian economies. Results of the analysis reveal a mechanism of efficient contract enforcement based on personal ties in small rural communities and identify weakening of the community relationship as a major factor underlying a shift from the share to the fixed-rental contract. The field survey of this study was conducted in 1982.

8.1 Jeepney Operations and Hypotheses

Anyone who has ever visited the Philippines will not be able to forget a swarm of decorative street vehicles coloured in red, yellow, and silver, busily carrying passengers. This minibus with 14–16 seats is called a 'jeepney' because it was originally converted from the US military jeep after World War II.

Mode of operations

The operations of jeepneys are typical of the 'information' sector in developing countries in contrast to the 'formal, incorporated' sector.[2] Jeepneys are owned by private owners who are mostly *petit bourgeois* in local communities, such as shopkeepers, wealthy peasants, retired schoolteachers, and government officials. Most own only one or two vehicles. Jeepneys are operated either by the owners or their family members or by drivers who rent the vehicles on a daily basis.

Jeepney operations are very flexible. They usually run along a designated route from one depot to another, picking up and dropping off customers anywhere on the way for a posted fare for a certain distance. However, short detours are easily made for an additional payment. Jeepneys can also be hired for private use by negotiation.

To operate jeepneys for public transport, owners must secure a special licence plate labelled PUV (public utility vehicle) from the Board of Transportation. However, there are many jeepneys that operate without these plates, especially in rural areas, partly because it is costly to secure them and partly because jeepneys that bear these plates are limited to designated routes, whereas vehicles without them can run anywhere, even though they are subject to the risk of being caught by occasional police inspections. The jeepney fare is determined by the government, and this fare is surprisingly well observed by jeepney operators, despite apparent weakness in the government enforcement apparatus.

Usually jeepney-owners and drivers on the same route form an association. The owners pay an initial admission and annual membership fee per vehicle, and the drivers contribute daily dues. These associations employ dispatchers to co-ordinate operations at depots and pay compensation to members in cases of sickness and accidents, and as penalty (or bribery) payments to police. However, the associations do not seem to function as a barrier to entry into jeepney operations. Not only is membership granted easily but there are also many non-member jeepneys operating. The associations do not specially attempt to block operations of non-member jeepneys on their routes, but non-members are not allowed to stop at the depots at the terminals. Nor do the associations have any function in mediating relations between jeepney-owners and drivers. On the whole, the relations between owners and drivers are competitive, with no decisive monopoly or monopsony power on either side.

Rental contracts

There are two forms of jeepney rental contract. One is the fixed-rental contract in which a driver pays a fixed sum of money to an owner for the use of a jeepney for a day.[3] Another is the share contract, in which both revenue and expenses are shared by an owner and a driver; typically, the driver pays the operating costs of fuel and oil (and, in some cases, minor repairs such as fixing a flat tyre) as well as the costs of snacks, beverages, and cigarettes from the day's revenue and takes the residue to the owner's house at the end of a day's operation; from this the owner gets two-thirds and the driver one-third. The deduction of non-operating expenses implies that two-thirds of those expenses accrue to the driver as a payment in kind. Since this payment is independent of the jeepney's revenue, the share contract amounts to a linear payment system of our basic model (Chapter 2) in which the driver receives a fixed fee plus a portion of the net revenue. There is no case, at least in our observation, in which the driver is employed by the owner at a fixed wage-rate.

An obvious question is: how does the owner ensure that the driver observes the terms of the share contract? There is no formal mechanism of checking the jeepney's earnings. The driving distance can be metered roughly by the fuel consumption, which can be ascertained from a petrol-station receipt, but there is no way of knowing how many passengers the jeepney carried during the day. It would be even more difficult to enforce contractual terms on the level of work-effort and non-operating expenses if such terms were stipulated. What mechanism prevents drivers from cheating and shirking so as to make the share contract viable is a major question to be investigated in this chapter.

Almost all contracts, both share and fixed-rental, are made in informal, oral forms. Although the contract is on a daily basis,[4] it usually continues to be renewed. However, it can be terminated at any moment according to the will of either party. The determination of contract terms is left free to bilateral negotiations between owner and driver with no government regulations nor social constraint. The fact that the fixed-wage labour contract is not adopted for jeepney operations corresponds to the tendency that the fixed-wage contract is dominated by the share or the fixed-rent land-tenancy contracts unless land tenancy is socially and legally prohibited (Chapters 4 and 6).

Interestingly, there is a distinct pattern in the interregional distribution of share and fixed-rental contracts. Using an extensive survey conducted over the Southern Tagalog region that encompasses Greater Manila and the provinces of Batangas, Cavite, Laguna, Rizal, and Quezon, we are able to map the distribution of jeepney routes by the form of rental contract adopted (Fig. 8.1). It is clear that the fixed-rental contract is adopted mainly in Greater Manila (shaded in Fig. 8.1) and neighbouring urbanized areas and that the share contract is common in rural areas away from Manila. In the Greater Manila area, not a single case of share contract was observed.

Fig. 8.1 represents a cross-section image of the historical change from a community-based institution (for example, the share contract) to a market-based institution (for example, the fixed-rent leasehold contract) in the process of urban industrial development.

Alternative hypotheses

In this section we attempt to specify alternative hypotheses to explain the peculiar distribution of jeepney rental contracts, given the obvious difficulty in enforcing the share contract. One possible hypothesis, whether plausible or not, is to assume Marshallian inefficiency together with patrons' (owners') altruistic behaviour in guaranteeing minimum subsistence to clients (drivers) in rural communities. As discussed in Chapters 2 and 3, the Marshallian theory is partial in the sense that only the optimization of tenants is considered. If owners' utility maximization is considered, the

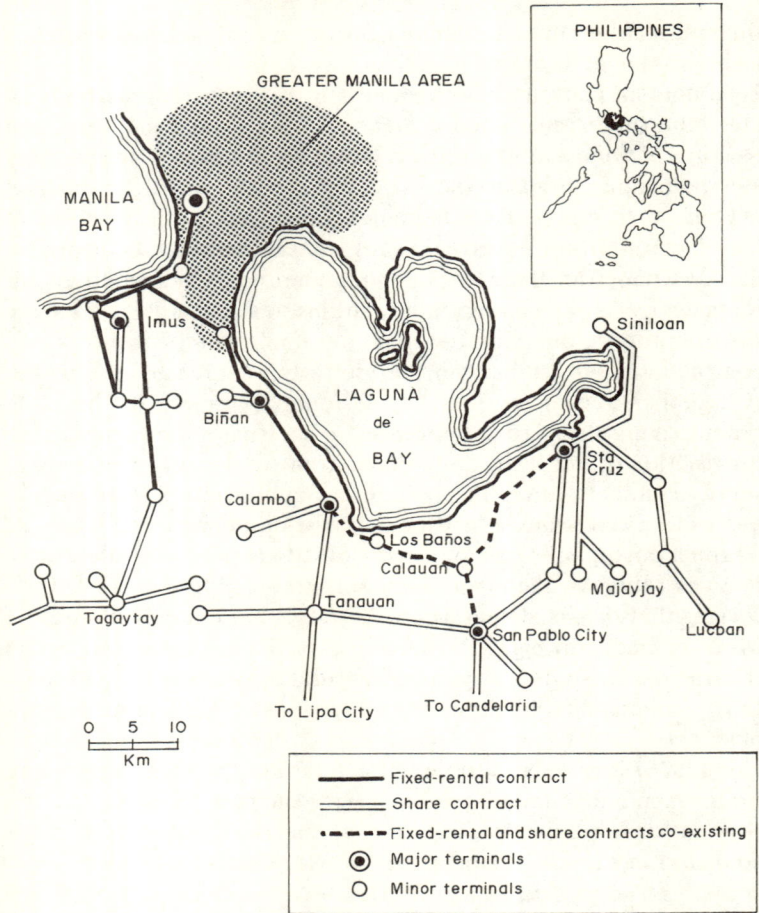

Fig. 8.1 Distribution of jeepneys by form of rental contract in the Southern Tagalog Region, Philippines, 1982

Marshallian equilibrium under share tenancy cannot exist because no owner will accept having the share rent lower than the fixed rent—in so far as both the share and fixed-rent contracts are available options, as is the case with jeepneys. The Marshallian theory does not explicitly consider uncertainty. In the world of uncertainty, the Marshallian theory is less likely to hold because owners should be able to collect a higher rent from share tenants for low risks than from leasehold tenants. The Marshallian solution is possible only on the assumption of patrons' altruistic behaviour in pre-capitalist rural communities where they reduce the risk of their poor clients' income declining below the minimum subsistence level at the expense of their own income (Section 3.1). This behaviour in pre-capitalist society is assumed in a large body of literature in political science and sociology from Karl Marx and Russian *Narodnik* to the recent 'moral economy' school (Mitrany 1951; Hayami 1990).

Once such behaviour is assumed, the peculiar distribution of jeepney rental contracts can easily be explained. As capitalist markets penetrate into rural communities, patron–client ties are replaced by impersonal market relations, and patrons, freed from the community obligation to guarantee minimum subsistence to poor neighbours, shift from the inefficient share contract to the more efficient and profitable fixed-rental contract. At the same time, drivers show less preference for the share contract because the risk associated with daily fluctuations in demand for jeepney services is likely to be higher in the isolated rural market than in the wide urban market in which a large number of the causes of demand fluctuation tend to cancel each other. This hypothesis may be called the 'altruist-community hypothesis'.

An alternative hypothesis corresponds to the case of enforceable contract under uncertainty (Section 3.3). This assumes the strong contract-enforcing mechanism of a small local community through long-term personal relations that prevents Marshallian inefficiency from occurring. The enforcement cost of share-rental contracts for jeepneys seems to be very high because drivers' input levels and output (revenue) levels are not easily checked by owners.[5] It is very easy for a driver to shirk or cheat his owner for several days. However, in a small community in which passengers and shopkeepers close to a depot are all friends of the owner, cheating will sooner or later be detected. The possible detection of dishonesty and the development of a bad reputation would involve a very high cost for a driver in a small community because he would not only lose his contract with the present lessor but also find it difficult to rent a jeepney from other owners (Chapter 4). The closer the patron–client relationship between owner and driver, the higher the penalty because the detection of one moral hazard would endanger the whole range of benefits from stable employment to emergency assistance (Chapter 5). Indeed, it is common for owners to advance interest-free loans to drivers in case of emergency.

While the drivers' opportunistic behaviour may be suppressed in such a community, risk is likely to be a serious concern for those engaged in jeepney operations which involve a relatively large outlay for fuel, and, if business is bad, can result in loss of income. It is difficult for tenant-operators to reduce risk by combining fixed-rent and fixed-wage contracts because the latter do not apply to jeepney operations and because jeepney operation requires specific skills not only in driving but also in such matters as appropriate timing in departure from the depot, which would make it less profitable for drivers to use part of their time for other wage employment.[6] Thus it is reasonable to expect that drivers prefer the share contract to the fixed-rental contract and that owners accept the share contract if they are offered a higher rent than under the fixed-rental contract.

In the course of urban industrial development, the cost of enforcing the share contract is increased as the personal relationships in small local communities are replaced by impersonal market relations; this, together with the lowered risk in a wider market, would result in a shift from the share to the fixed-rental contract. If this hypothesis is correct, Marshallian inefficiency cannot be observed for the share contract that is adopted where both the share and leasehold contracts are available options. This hypothesis may be called the 'efficient-community hypothesis'.

In between the altruist-community and the efficient-community hypotheses, there is another hypothesis that may be called the 'inefficient-community hypothesis'. This corresponds to the case of unenforceable contract under uncertainty (Section 3.3). Assuming the difficulty from the principals' (owners') point of view of monitoring agents' (drivers') inputs, this hypothesis predicts that the level of inputs and output under the share contract will be smaller than under the fixed-rental contract, while the rental rate is higher under the former than under the latter by a risk premium and an implicit penalty on the insufficient supply of work-effort, provided that the elasticity of substitution between land (capital) and work-effort is less than unity (equation 3.30). According to the results shown in the Appendix to this chapter, the elasticity of substitution is positive but significantly less than unity. Therefore a lower level of inputs under the share contract than under the fixed-rental contract seems inevitable unless tenants' input levels are effectively monitored by owners, as assumed in the efficient-community hypothesis. In terms of this theory, the shift from share to leasehold contract in the course of urban industrial development can be explained by the reduction in risk due to expanded markets. This explanation holds as long as drivers do not become less risk-averse compared with owners with the passage of time (Weitzman 1980).

Possible differences in the magnitudes of revenue (output), inputs, rental, and driver income from jeepney operations between share and leasehold contracts predicted from the three alternative hypotheses are summarized in Table 8.1.[7] As is clear from a comparison of this table with Table 3.1, the altruist-community hypothesis is equivalent to the case of unenforceable contract under certainty, the efficient-community hypothesis to that of enforceable contract under uncertainty, and the inefficient-community hypothesis to that of unenforceable contract under uncertainty with $\sigma < 1$.

TABLE 8.1. *Orderings in the magnitudes of output, input, rental, and driver income from jeepney operations between the share and fixed-rental contracts, predicted from the three alternative hypotheses*

Hypothesis	Output	Input	Rental	Driver income
Altruist-community	S < R[a]	S < R	S < R	S ≧ R
Inefficient-community	S < R	S < R	S > R	S < R
Efficient-community	S = R	S = R	S > R	S < R

[a] S and R refer, respectively, to the share and fixed-rental contracts.

8.2 Testing the Alternative Hypotheses

An extensive survey and its limitations

The first set of data for testing the alternative hypotheses was collected in an extensive survey to identify the regional distribution of forms of contract, as shown in Fig. 8.1. In the survey, we had a group interview with a number of drivers and dispatchers at each depot on forty jeepney routes. In addition to the question concerning the most frequent form of contract, we asked drivers what they considered their average, highest, and lowest revenues per day during the previous week. We attempted to measure risk by taking the ratio of the difference between the highest and lowest revenues to the average revenue. These data are averaged by form of contract (Table 8.2).

TABLE 8.2. *Average revenue, jeepney rental, and driver income per jeepney per day in a sample of jeepney routes classified by dominant form of contract in the Southern Tagalog region, Philippines*

Dominant form of contract (no. of routes)	Gross revenue (₱/day)	Jeepney rental (₱/day)	Driver income (₱/day)	Variation in gross revenue[a] (%)
Share (11)	120	46	34	59
	(40)	(18)	(10)	(31)
Coexisting (7)[b]	175	63	44	39
	(27)	(14)	(8)	(15)
Fixed-rental (22)	203	73	68	34
	(39)	(16)	(17)	(16)
t-statistics				
Share vs. coexisting	3.38*	2.35**	2.26*	1.69*
Coexisting vs. fixed	1.69	1.47	3.45**	0.64
Share vs. fixed	5.71**	4.41**	7.10**	2.56**

Note: Numbers in parentheses are standard deviations.

[a] Ratio of difference between a week's highest and lowest gross revenues to the average gross revenue.
[b] Both share and leasehold contracts coexist in more or less similar proportions.
* Difference is significant at the 5% level.
** Difference is significant at the 1% level.

The results show that average gross revenue, average rental to owners, and average income of drivers are all significantly lower for the share than for the fixed-rent contract, whereas variations in gross revenue are greater on the routes where the share contract was most common. Such results appear to support strongly the altruist-community hypothesis. However, these results are highly deceptive. In the urban areas where the fixed-rent contract prevails, the cost of living and opportunity wage-rates for drivers are much higher than they are in the rural areas. The working conditions are also much harsher under heavy traffic in the urban areas.

Thus the higher gross revenues and the higher incomes of drivers on the routes dominated by fixed-rental contracts do not necessarily

The Case of Upland Farming in Indonesia 137

reflect the inefficiency of the share contract relative to the fixed-rental contract. For a rigorous test, it is necessary to compare the data from the areas where the economic conditions, such as the opportunity wage-rate and the working conditions, are similar. On the other hand, our measure of the variations in gross revenue might reflect reasonably well interregional differences in risk associated with the two forms of contract, even though the data are admittedly very crude. This is consistent with the general finding of share-tenancy literature that the incidence of share contracts is positively associated with yield variation (Section 6.2).

Efficiency tests based on the intensive survey

To collect more appropriate data for testing the hypotheses, an intensive survey was conducted on the routes from Calamba to Los Baños and from Los Baños to San Pablo in the province of Laguna, where both the share and the fixed-rental contracts coexist (Fig. 8.1). The survey consisted of interviews with randomly selected drivers, owners, and repair shops. It is reasonable to assume that the economic conditions, such as risk and opportunity wage-rate, that surround drivers and owners are the same on the same route.

First, based on the survey of drivers, output and inputs per jeepney per day are compared among the three different tenure classes in order to test whether the share contract is subject to allocative inefficiency. The outputs are measured in terms of both gross revenue and gross value added after deducting fuel and oil costs from the gross revenue; the inputs are represented by diesel consumption and drivers' working hours (Table 8.3). The results show that both input and output levels under the share contract are not significantly smaller than those under the fixed-rental contract, which supports the efficient-community hypothesis relative to the alternative two hypotheses. The results of equal allocative efficiency between the share and the fixed-rental contracts are consistent with many empirical studies of share-cropping under no constraint on contract choice, as reviewed in Chapter 6.[8]

A major anomaly in the results presented in Table 8.3 is that the revenue and diesel consumption are lower for owner-operators than for tenant-operators, whereas the length of working hours is about the same. One possible explanation is to assume a stronger preference for leisure by relatively richer owner-operators; thereby

TABLE 8.3. *Average revenue, value added, labour, and current inputs per jeepney per day by form of contract on the Calamba San Pablo route, Laguna, Philippines*

Form of contract	Gross revenue (₱/day)	Gross value added (₱/day)[a]	Diesel consumption (₱/day)	Working hours (hours/day)
Share	202	120	73	12.1
	(32)	(22)	(14)	(1.6)
Fixed-rental	207	124	76	12.3
	(38)	(20)	(22)	(1.7)
Owner	184	109	67	12.0
	(34)	(23)	(13)	(1.9)
t-statistics				
Share vs. fixed	0.61	0.82	0.56	0.54
Share vs. owner	2.29*	1.95	2.04*	0.30
Fixed vs. owner	2.26*	2.43*	1.74	0.66

Note: Based on a survey of 52 share, 25 leasehold, and 24 owner-drivers.
Numbers in parentheses are standard deviations.

[a] Gross revenue minus diesel and oil costs and daily contribution to the drivers' associations.

* Difference is significant at the 5% level.

they are resting longer during working hours. (Note that 'working hours' here is defined as the number of hours from the time that drivers begin working to the time that they finish.) Another possible explanation is that the owner-operators are more careful in operating their own vehicles and spend more time on maintenance.[9] In fact, it is a common opinion that fixed-rental operators are the most reckless drivers and owner-operators are the most careful. To verify this common belief, data were collected from repair shops and jeepney-owners as to the costs of maintenance and depreciation. Several indicators in Table 8.4 show that the capital depreciation rate and the maintenance cost are the highest for fixed-rent operators and the lowest for owner-operators; these results are consistent with expectations from the land-quality transaction-cost model of share tenancy advanced by Murrell (1983) and others (Section 3.2).

The results in Table 8.4 suggest that the comparisons of output

TABLE 8.4. *Estimation of average depreciation and repair costs of jeepneys by form of contract on the Calamba–San Pablo route, Laguna, Philippines*

Form of contract	Durability of new tyres (months)[a]	Engine overhaul interval (years)[a]	Usable life of jeepney (years)[b]	Depreciation and maintenance cost (₱/year)[b]
Share	7.6	3.2	16.9	7,298
	(0.9)	(0.6)	(3.3)	(879)
Fixed-rental	6.5	2.4	14.4	8,668
	(0.8)	(0.5)	(4.4)	(1,625)
Owner	11.4	4.2	17.8	6,281
	(1.0)	(0.7)	(3.3)	(1,975)
t-statistics				
Share vs. fixed	2.64*	2.74*	1.24	2.00*
Share vs. owner	7.73**	2.72*	0.65	1.32
Fixed vs. owner	10.52**	5.48*	2.39*	3.25**

Note: Numbers in parentheses are standard deviations.
[a] Based on a survey of 7 repair shops.
[b] Based on a survey of 7 share owners, 9 fixed-rental owners, and 24 owner-operators. Average repair cost is obtained from a survey of owners, and depreciation of jeepneys is calculated by assuming a linear depreciation schedule and zero salvage value.
* Difference is significant at the 5% level.
** Difference is significant at the 1% level.

levels made in Table 8.3 must be adjusted for differences in the depreciation and maintenance cost. Such adjustments are made in Table 8.5 (cols. 1–3).[10] The results show that differences in the net value added among three tenure classes are much smaller than in the gross-value added and are statistically insignificant at conventional levels. Thus, the hypothesis on equal efficiency between owner- and tenant-operators is not rejected, even though the possibility of stronger leisure preference by owner-operators cannot be ruled out.

Table 8.5 (cols. 4–6) also compares the income distribution between owners and drivers under the share and fixed-rental contracts. It is shown that the net rental of jeepneys after deducting capital depreciation and maintenance costs is significantly higher

TABLE 8.5. *Distribution of value added per day from jeepney operation among owners and drivers by form of contract on the Calamba–San Pablo route, Laguna, Philippines*

Form of contract	Gross value added	Depreciation and maintenance cost[a]	Net value added[b]	Gross rental	Net rental[c]	Driver income[d]
	(1)	(2)	(3)	(4)	(5)	(6)
Share	120	24	96	71	47	49
	(22)	(3)	(25)	(13)	(17)	(10)
Fixed-rental	124	29	95	68	39	56
	(20)	(5)	(25)	(12)	(17)	(14)
Owner	109	21	88	—	—	—
	(23)	(7)	(32)			
t-statistics						
Share vs. fixed	0.82	2.00	0.06	1.11	2.17**	2.78**
Share vs. owner	1.95	1.32	1.41	—	—	—
Fixed vs. owner	2.43*	3.25**	1.22	—	—	—

Note: Numbers in parentheses are standard deviations.

[a] Converted from the annual cost in Table 8.4, assuming 300 days of operation per year.
[b] (3) = (1) − (2).
[c] (5) = (4) − (2).
[d] (6) = (3) − (5).
* Difference is significant at the 5% level.
** Difference is significant at the 1% level.

for the share contract,[11] whereas the drivers' income is significantly higher for the fixed-rental contract. This finding is prima-facie evidence for the drivers' risk-aversion under uncertainty. Altogether, the results in Tables 8.3 to 8.5 support the efficient-community hypothesis, which predicts that the share contract is subject to no allocative inefficiency but only to an income distribution between owners and tenants different from that of the fixed-rental contract.

8.3 Conditions of Effective Contract Enforcement

How are the terms of a share contract enforced effectively so that Marshallian inefficiency does not occur? Our hypothesis advanced in Chapters 4 and 5 is that close social interactions and enduring personal ties among owners and tenants in small rural communities make it relatively easy for owners to detect tenants' dishonest behaviour and to demand high penalties from tenants who are discovered shirking or cheating.

If our hypothesis is correct, personal ties among owners and drivers should be stronger under the share than the fixed-rental contract. This expectation is confirmed by the finding that a significantly higher proportion of jeepney-owners under the share contract were the relatives of tenant-drivers (Table 8.6, col. 2). Furthermore, the residence stability of drivers, measured by the ratio of the number of years living in the present residence to age, is significantly higher for the drivers under the share contract than under the fixed-rental contract (Table 8.6, col. 4), indicating that the share-drivers are living much longer in the same community than the fixed-rental drivers, thereby developing closer personal relations with the owners as long-time neighbours. The penalty of losing multi-stranded personal ties with an old patron and developing a bad reputation within a small circle of relatives and close neighbours would be high so that tenants would be forced to refrain from shirking and cheating. The efficient contract-enforcement mechanism of a small community observed in the previous chapter for agricultural land tenancy in Indonesia is thus found to be applicable to jeepney rental contracts in the Philippines.

It seems reasonable to expect that the enforcement of implicit contracts based on such personal ties is subject to scale diseconomies.

TABLE 8.6. Socio-economic characteristics of jeepney-owners and drivers by form of contract on the Calamba–San Pablo route, Laguna, Philippines

Form of contract	No. of jeepneys owned(%)	Ratio of relatives (%)	Ratio of mechanics and/or drivers (%)	Residence stability index (%)[a]	Age (years)	Professional driving experience (years)
	(1)	(2)	(3)	(4)	(5)	(6)
Share	1.8	54	33	87	32	11
	(1.1)			(27)	(7)	(7)
Fixed-rental	3.2	28	56	64	34	12
	(3.3)			(39)	(9)	(9)
Owner	1.1	—	—	76	39	14
	(0.3)			(32)	(10)	(10)
t-statistics						
Share vs. fixed	2.68**	2.13*[b]	1.95*[b]	2.97**	0.95	0.82
Share vs. owner	3.20**	—	—	1.53	3.28**	1.84*
Fixed vs. owner	3.11**	—	—	1.15	1.81*	0.79

Note: Numbers in parentheses are standard deviations.

[a] Ratio of the number of years living in the present residence to age.
[b] z (standard normal)-statistics
* Difference is significant at the 5% level.
** Difference is significant at the 1% level.

The Case of Upland Farming in Indonesia 143

This is confirmed by the finding that the average number of jeepneys per owner was significantly smaller under the share than it was under the fixed-rental contract (Table 8.6, col. 1). Since it is reasonable to expect that large owners are less risk-averse than small owners, we would expect owners who have share contracts to own a larger number of jeepneys than those who have fixed-rental contracts if the risk-attitude is decisive in contract choice. The fact that the average number of jeepneys per owner is larger under the fixed-rental contract suggests that the cost to an owner of share-contract enforcement will rise sharply as the scale exceeds a limit manageable by personal interactions; beyond this limit, owners would be forced to adopt the fixed-rental contract that involves a smaller enforcement cost at the expense of higher rental from the share contract.

However, the enforcement of a fixed-rental contract is not costless. A major problem for lessors with fixed-rental contracts is how to prevent the reckless use of jeepney by lessees—a problem similar to that of abuse of land by tenants in the land-quality transaction model of share tenancy. It is interesting to observe that the percentage of owners who are mechanics and/or professional drivers is higher under the fixed-rental contract than under the share contract (Table 8.6, col. 3). This finding suggests that those who have the ability to detect undue capital breakdown and depreciation have a comparative advantage in enforcing fixed-rental contracts.

A hypothesis advanced by Reid (1973, 1976b, 1979b) on the life-cycle 'agricultural ladder' from agricultural labourer, to sharecropper, to fixed-renter, to owner-operator, and finally to non-cultivating landlord does not seem to apply to the jeepney case because the proportion of owners who are in a position to give advice to drivers, such as mechanics and professional drivers, is smaller for the share contract; however, there is some indication that the life-cycle ladder from the share to the fixed-rental contract, and then to the owner-operator does exist in terms of age and professional driving experience (Table 8.6, cols. 5 and 6).

8.4 Conclusion

The peculiar distribution of jeepney rental contracts in the Philippines—the fixed-rental contract is common in the urban metropolis, and the share contract is common in rural areas—has suggested three alternative hypotheses. The results of an intensive survey over the route where both share and fixed-rental contracts coexist are consistent with the efficient-community hypothesis in which resource allocation in terms of input and output levels is equally efficient between the share and fixed-rental contract and only income distribution differs due to different risk-sharing involved in the share and fixed-rental contract. In terms of this hypothesis, the shift from the share to the fixed-rental contract corresponding to urban industrial development is explained both by the reduction in risk due to the formation of a wider and more stable market and by the increase in transaction costs associated with the share contract as the close and stable personal ties in rural communities are undermined by urbanization.

These results imply that if both risk premium and transaction costs are counted in the evaluation of economic welfare, the welfare levels of both owner and tenant as well as total economic efficiency is higher for the share contract under the conditions of small rural communities than those for the fixed-rental contract, and vice versa under the conditions of large urban societies. Government regulations on the form of contracts (such as the prohibition of share tenancy in Philippine land-reform laws and the allegedly planned prohibition of fixed-rental contracts in the operation of jeepneys in order to prevent reckless driving in metropolitan Manila), without due consideration of specific socio-economic environments, may impair both efficiency and equity.

It should be clearly understood that no contract form is universally efficient. Relative efficiencies of various contracts depend on the social and economic conditions that determine risk and transaction costs. As this study of an informal sector of the Philippines confirms, common people in developing economies seem to have sufficient ability to change the form of contract so as to be efficient under changing environments in the course of economic development.

Appendix: Estimation of Elasticity of Substitution

In order to perform a rigorous test of contractual efficiency, it is attempted to estimate the elasticity of substitution (σ) à la Arrow et al. (1961). The basic models used for estimation are

The Case of Upland Farming in Indonesia

and

$$\log (Q/L) = a + b \log W^* + c \log E + d \log E'$$

$$\log (Q'/L) = a' + b' \log W^* + c' \log E + d' \log E',$$

where Q and Q' are respectively gross revenue and value added from the jeepney operation per day; L is labour hours; W^* is the implicit wage-rate (driver income per hour); E is the age of the jeepney as a variable representing the quality of capital; and E' is the driver's driving experience in years as a variable representing the quality of driver's labour. It is well known that under competitive factor-market equilibrium b and b' measure the elasticity of substitution (between labour and the aggregate of other inputs, including current inputs such as fuel in the case of b; and between labour and capital in the case of b').

Since the share-drivers may undersupply work-effort, the equilibrium under share contract may be quite different from the competitive equilibrium. Therefore, the estimated coefficients of b and b' will be biased estimates of the elasticity of substitution if the implicit wage-rate of share-driver is used in estimation of the above equations. On the other hand, the implicit wage-rate of owner-operator is difficult to estimate because the observable income of owner-operators includes not only the return on his labour but also the return on capital (jeepney). Meaningful estimates of the elasticity of substitution can be obtained only in the case of the fixed-rental contract, since it can be reasonably assumed that drivers under this contract will not shirk. Therefore, the estimation is based on the sample of twenty-five fixed-rental drivers.

The estimated results are reported in Table 8.7. It is clearly shown that the estimates of elasticity of substitution are all significantly greater than zero but less than unity at the 1 per cent level for various specifications of the regression equation. Moreover, they are consistently clustered around 0.5. Such result is expected if capital and fuel are highly complementary. These results justify our assumptions that $0 < \sigma < 1$.

Notes to Chapter 8

1. This chapter draws heavily on Otsuka et al. (1986).
2. Informal operations of small-scale public transport, like jeepney operations, that use either automobiles or animal-drawn wagons are common in developing countries.
3. Though seldom happening, the rental is reduced when the revenue is extremely low—the rate of reduction being 10–20% of the normal rent.

TABLE 8.7. Estimates of the elasticity of substitution function

Regression number	Dependent variable	Coefficients of			Intercept	R^2
		Wage-rate	Age of jeepney	Driver's experience		
(1)	Q/L	0.440 (0.087)[a]			2.154	0.506
(2)	Q/L	0.503 (0.099)	−0.030 (0.029)	−0.025 (0.030)	2.151	0.507
(3)	Q'/L	0.502 (0.060)			1.554	0.739
(4)	Q'/L	0.546 (0.070)	−0.025 (0.020)	−0.006 (0.021)	1.532	0.736

[a] Numbers in parentheses are standard errors.

The Case of Upland Farming in Indonesia   147

4. There are exceptional cases in which the fixed-rental contract is made on a weekly basis.
5. Russel (1985) documents that share contract in taxicab operation in New York became unworkable as drivers did not turn on the taximeter. In the case of the jeepney, even the meter is not installed.
6. Jeepney-drivers' average daily income is comparable with the wage-rates of skilled craftsmen, such as carpenters and plumbers, which are two to four times higher than agricultural wage-rates.
7. The same sets of ordering that are predicted by the efficient-community and the inefficient-community hypotheses may also be predicted from a model of 'semifeudal exploitation' à la Bhaduri (1973), in which a landlord (owner) is a reckless monopsonist in labour demand and a monopolist in the supply of land (jeepney) and credit in an isolated local community. However, this is very far from the world of Philippine rural communities in which jeepneys are being operated. The equilibrium resource allocation in this model corresponds to that of the efficient-community model if the landlord can enforce contracts costlessly in the manner of a discriminating monopsonist. The difference between our model and the semifeudal exploitation model lies in the distribution of income between principal and agent, even though predicted orderings remain the same, as shown in Table 8.1.
8. This is also consistent with a recent study of Otsuka and Murakami (1989) who found equal efficiency between the share and the fixed-rental contracts in taxicab operation in Japan. According to their estimate, the elasticity of substitution between capital and labour is significantly less than unity.
9. Still another explanation is that the owner-operators are more risk-averse than the tenant-drivers. This hypothesis is unlikely because the owners' wealth position is higher than the tenants'.
10. Standard deviations of differences between two means calculated from different data sources, shown in cols 3, 5, and 6 of Table 8.5, are estimated from the weighted regression with binary explanatory variables. For methodology, see Kmenta (1971: 366 and 409–18).
11. This relation was confirmed from the survey results of seven share and nine fixed-rental owners; the average net share rental, ₱48 per day, is significantly greater than the fixed-rental, ₱37, at the 1% level.

9

Land Reform, New Technology, and Agrarian Contracts
The Case of a Philippine Rice Bowl

THE previous two chapters analysed the cases in which no institutional constraint limited the scope of choice among labour-employment and land- (or capital-) rental contracts. In neither case was the presence of the permanent-labour contract reported. This situation is consistent with our theoretical prediction (Chapter 4) that the permanent-labour contract tends to be dominated by the land-tenancy contract if it is difficult for landed principals to monitor their agents' work-effort. It is also consistent with a finding from the global survey of empirical studies (Chapter 6) that the use of permanent labour is associated with socio-political regulations to prohibit the practice of land-tenancy contracts.

If landowning principals are forced to employ permanent labour because of institutional constraints against land tenancy, resulting resource allocation is likely to be inefficient, as is the case with share tenancy under the constraint against fixed-rent tenancy. In this chapter we attempt to show how land-reform regulations on tenancy contracts have resulted in the diffusion of permanent-labour contracts in a major rice-producing area in the Philippines. We also try to examine the efficiency of resource allocation under permanent-labour contracts.

9.1 Study Site and Data Collection

The site chosen for this analysis is Central Luzon, popularly known as the rice bowl in the Philippines. Lowland areas in Central Luzon are mostly irrigated and double-cropped for rice. As in other rice-producing areas in the Philippines, landlord–tenant relations are pervasive. While the old-settled coastal areas of Central Luzon have traditionally been characterized by scattered landholdings by relatively small landlords, in the landlocked inner part, especially the Province of Nueva Ecija, large rice haciendas were developed in the nineteenth century under Spanish rule (Hayami and Kikuchi 1982, ch. 4).

Unlike the plantations of export cash-crops based on hired labour under centralized management, the rice haciendas were usually operated by small share-tenants in a decentralized manner. The share-croppers' operations, especially harvesting and threshing, were supervised strictly by the hacienda management consisting of a farm manager (*encargado*) and overseer (*katiwala*), while the owner usually lived in Manila (Hayami and Kikuchi 1982, ch. 4). They were often heavily indebted to *hacienderos* so that the subtraction of rent and debt repayment from the harvest left often less than sufficient for subsistence until the next harvest with no escape from perpetual indebtedness. The sharp class confrontation between *hacienderos* and share-croppers had made this region the nursery of agrarian unrest, as represented by the Hukbalahap (or Huk) revolt during and after World War II. For the compelling need of government to appease agrarian unrest, the major effort of land reform in the Philippines from the first president Manuel Quezon (1935–9) to Ferdinand Marcos (1965–86) was concentrated on this region (Hayami *et al.* 1990). Therefore, the impact of land-reform programmes on agrarian contract relations can be most clearly observed there.

This case-study relies on three surveys conducted at the International Rice Research Institute (IRRI), Agricultural Economics Department. The first is the Central Luzon Loop Surveys conducted periodically from 1966/7 to 1986/7. This survey covered fifty-eight to 149 rice farmers along the highway loop across five provinces in Central Luzon with a standard questionnaire mainly on rice production practices (Herdt 1987). The second is the Permanent Labour Contract Survey conducted in 1987 covering thirty-six municipalities in Central Luzon. In this survey farmers or groups of farmers met randomly along the survey trip route were interviewed with a simple questionnaire about the scope and the nature of permanent-labour arrangements in their surroundings (Hayami and Otsuka 1992). The third is the Nueva Ecija Village Survey conducted in 1989 (Otsuka *et al.* 1989, 1993). This survey presents a highly intensive data collection in one village in the Nueva Ecija province (henceforth called the Nueva Ecija village), covering seventy-one farmer households and forty-eight landless-labourer households;

this sample survey was preceded by a baseline survey in 1987 for making the sampling framework. While the extensive surveys like the Central Luzon Loop Survey and the Permanent Labour Contract Survey are effective in identifying broad changes in production relations, investigation into elusive and sensitive issues on the choice of contracts often violating land-reform regulations must rely on the intensive survey at a village level like the Nueva Ecija Village Survey.

9.2 Changes in Labour Relations

First, we outline the emergence of a new form of permanent-labour contract in irrigated rice areas in Central Luzon.

Traditional permanent labour

Rice-farming in the Philippines is known for its high dependence on hired labour (Barker and Cordova 1978), but the hired-labour contract used has usually been short-term, mostly for a day or for a specific task which can be accomplished within a day. Permanent labourers employed for a year or a crop season, relatively common in India and Nepal, have seldom been found in the rice sector of the Philippines or other countries in South-East Asia, although they are common in the plantations of cash crops, e.g. *dumaan* in sugar haciendas in the Island of Negros. Indeed, sociological and economic investigations into the agrarian structure of rice villages in the Philippines in the past rarely reported a significant incidence of permanent-labour contracts (Hester and Mabun 1924; Rivera and McMillan 1954; Anderson 1964; Takahashi 1969; Umehara 1974; Hayami *et al.* 1978; Hayami and Kikuchi 1982; Ledesma 1982).

However, there is a historical study recording the practice of a permanent-labour contract among Ilocano settlers who migrated from the tobacco-producing region in the north to Central Luzon when it opened up for rice production in the late nineteenth century (McLennan 1982). In this agreement, called *kasugpong* (an Ilocano word meaning 'helper') contract, a permanent labourer is a young single male, lives in a farmer-employer's house with food and clothing provided, and works exclusively at the employer's farm helping in operations of all kinds. In sparsely populated frontier lands where peace and order had not been established (e.g. high incidence of theft of draught animals), settlers would have found it convenient to have young male helpers living together on their farms. At the same time, this farm-helper arrangement would have provided the young men with a first step on the 'agricultural ladder' (Spillman 1919) from agricultural wage-worker to tenant-farmer, as it enabled them to acquire farming skills and savings for such purchases as draught animals. However, this arrangement was not so commonly practised among Tagalog people who began settlements in Central Luzon from the south much earlier than the nineteenth century.

New permanent labour

Recently a new form of permanent-labour contract, similar to 'semi-attached labour' in India, (Rudra 1971; Ghose 1980; Bardhan and Rudra 1981; Basant 1984), has been spreading in Central Luzon. In this contract, a permanent labourer can be either single or married with family. He and his family live in a shanty either inside or outside the employer's residential quarters He has the obligation to perform certain preassigned tasks, while he is allowed to work outside his master's farm as a casual worker in order to supplement his income, paid in a fixed amount of paddy ranging from 10 to 30 cavans (1 cavan = 50 kg.), depending on experience and skill as well as the scope of tasks assigned to him, or a share of output (usually 10 per cent).

According to the Permanent Labour Contract Survey, the traditional farm-helper arrangements were found to be concentrated in a rainfed area lying from the Nueva Ecija–Tarlac border to the southern part of Pangasinan (Fig. 9.1). According to survey respondents, the farm-helper is an old institution in this area and existed before World War II, but it has not increased recently but has rather decreased because young workers have tended to migrate out to irrigated areas. In contrast, the new, semi-attached labourers were commonly observed in the irrigated areas of Nueva Ecija, Bulacan, and Pampanga. In these areas, we gathered that the new system has become common since the late 1970s or early 1980s.

Although the Permanent Labour Contract Survey in 1987 was not able to measure the rates of diffusion of the new permanent-labour

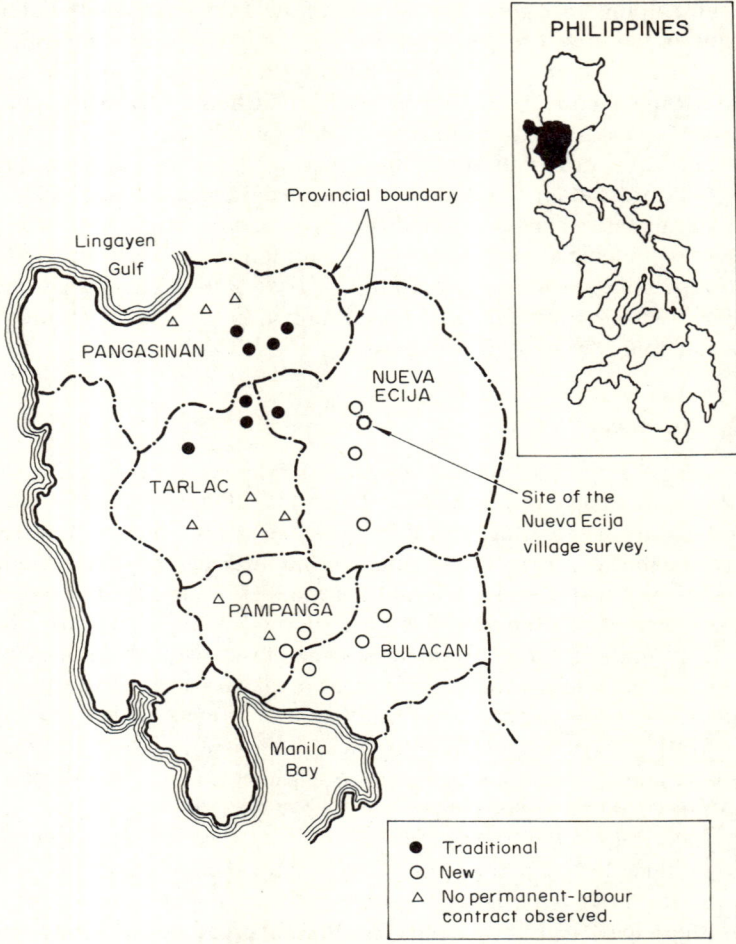

Fig. 9.1 Geographical distribution of different types of permanent-labour arrangements in twenty-eight municipalities, Central Luzon, August 1987 (based on the Permanent-Labour Contract Survey)

contract, the Central Luzon Loop Surveys conducted periodically since 1966 show that permanent labour as measured by the percentage of farmer respondents who employ permanent labourers increased from only about 5 per cent for 1966–70 to 24 per cent in 1986–7 (Table 9.1). Although the Loop Survey data do not classify

TABLE 9.1. *Changes in rice technology, land-tenure distribution, and the incidence of permanent labour in Central Luzon and the Nueva Ecija village (selected years)*

	Central Luzon			Nueva Ecija	
	1966	1970	1986	1970	1987
MV adoption rate (%)[a]	0	63	95	8	100
Double-cropping ratio (%)[b]	19	21	50	18	100
Rice yield (tonne/ha.)[c]	2.0	2.3	4.0	2.6	4.9
Land tenure distribution:[d]					
Owner	12	10	3	5	11
Share	75	55	16	80	0
Leasehold/CLT	13	35	65	15	73
Pawn	0	0	2	0	15
Mixed	0	0	14	—	—
TOTAL	100	100	100	100	100
Incidence of permanent labour[e]	5	6	24	15	36
Percent of landless labour household[f]	n.a.	n.a.	n.a.	13	22

[a] Percentage of parcels planted in modern varieties.
[b] Area with two rice crops per year.
[c] Average for wet and dry seasons.
[d] Distribution of farmers for Central Luzon, and distribution of land plots for the Nueva Ecija village.
[e] Percentage of farmers employing permanent labourers to the total number of farmers for Central Luzon, and percentage of permanent labourers to the total number of landless labourers for the Nueva Ecija village.
[f] Percentage to the total number of households in the village.
n.a.: not available

Sources: Central Luzon data based on the Central Luzon Loop Survey; Nueva Ecija village data based on the baseline survey for the Nueva Ecija Village Survey.

permanent labour between the traditional and the new forms, it is reasonable to assume that most of the increase has resulted from the increase in the new category. In fact, it is more likely that traditional farm-helpers decreased not only relatively but also in absolute number. The increase in permanent-labour contracts is also confirmed by the data from the Nueva Ecija Village Survey that show an increase in the proportion of permanent labourers in the number of landless agricultural workers from 15 per cent in 1970 to 36 per cent in 1987 (Table 9.1); in this village all the permanent labourers in 1987 were of the semi-attached labour type.

9.3 Conditions of New Permanent-Labour Contracts

Why did the new permanent-labour contract begin to spread rather suddenly in recent years over the irrigated areas of Nueva Ecija and Bulacan? The three major factors that appear to underlie this development are: implementation of land-reform programmes, development of new rice technology, and transfer of land-cultivation rights to urban residents.

Land reform

Land reform in the Philippines has traditionally consisted of two programmes, i.e. a shift from share to leasehold tenancy with a government-controlled fixed rent, and a redistribution of tenanted rice (and corn) land above a landlord's retention limit to tenants cultivating the land. These programmes had been implemented on the basis of the 1963 Agricultural Land Reform Code but until 1972 they were adopted only in the pilot project area in Nueva Ecija (de los Reyes 1972). The Code was amended in 1971 with automatic conversion of all share-tenants to leaseholders. The 1971 Code was strongly enforced by Presidential Decrees Nos. 2 and 27 under Martial Law proclaimed in 1972. The landlord's retention limit was successively reduced from 75 to 7 ha. (Hayami *et al*. 1990, ch. 3). Progress in the land-reform programmes is reflected in a marked decline in share tenancy which was replaced by either leasehold tenancy or the holding of CLTs (Certificates of Land Transfer) that entail the holder to cultivate the land while paying amortization to the government (the Land Bank) before obtaining possession,[1] as shown in the data of both the Central Luzon Loop Survey and the Nueva Ecija Village Survey (Table 9.1). Both land rent on leasehold land and amortization payments by CLT-holders are fixed at about 25 per cent of average rice yield for three normal years preceding the year of programme operation.[2]

New rice technology

Since the implementation of land-reform programmes, dramatic increases in rice yields have occurred. With accelerated investment in irrigation systems since the 1950s, many of the rice areas in Nueva Ecija and Bulacan were irrigated and shifted from single- to double-cropping in the 1970s, especially after the completion of Pantabangan Dam in 1974. According to the Loop Survey, double-cropped area increased from about 20 per cent of total rice area in 1970 to 50 per cent in 1986. Based on the improved irrigation infrastructure, sharp increases in rice yields have resulted with the diffusion of modern rice varieties (MVs) developed by IRRI and the increased application of fertilizer and chemicals as shown in the data of Table 9.1. Typically, paddy yields almost doubled from about 2 tonnes in the late 1960s to a level of 4 to 5 tonnes per hectare since the mid-1980s.

Increased demand for labour during the peak season resulted in an increase in hired labour relative to family labour (Barker and Herdt 1985; Herdt 1987). The substitution of hired for family labour was further encouraged by the increased income of land-reform beneficiaries. With both land rents and amortization payments being fixed at the time of land-reform operations, the yield increases have resulted in major gains to land-reform beneficiaries, i.e. leaseholders and amortizing owners (Otsuka 1991). As their incomes rose, disutility of their labour increased. Some of them have started non-farm businesses such as small trades and manufacturing based on capital saved from increased farm incomes. Naturally, the preference for leisure and urban occupations has become greater for their children who have received better education than their parents. Thus high demands have developed to substitute hired labour for family labour in rice-farming operations.

Traditionally, the transplanting and harvesting of rice were performed by hired labour. The former was contracted out on a daily-wage basis to a crew of transplanters organized by a labour-drafter called *kabisilya*, while harvesting was usually carried out through a contract called *hunusan* by which those who participated received a certain share of the harvested crop. As the need for weeding increased with the introduction of short-statured MVs, hired labour for this purpose increased either on a daily-wage basis or through a new contractual arrangement called *gama* under which workers weeded without pay but had an exclusive right to harvest a particular plot or plots and receive their share of harvests (Hayami and Kikuchi 1982, chs. 4 and 5).

Land preparation had traditionally been the major task that farmers were expected to perform themselves because it calls for skill in ploughing and harrowing without harm to their draught animals. Recently, however, ploughing has increasingly been contracted out to custom services of power-tillers (small tractors). The remaining tasks for farm-operators and family members are those requiring care and judgement and, therefore, are more difficult to monitor, such as water control, and fertilizer and chemical applications. The relative importance of these tasks has increased with the development of irrigation systems and the diffusion of new rice technology. Unlike the labour for transplanting and harvesting, the results of labour applied to these tasks are difficult to observe, while the demand for such labour is spread thinly over a crop season in an unpredictable manner. Therefore transaction costs for casual labour for these tasks tend to be prohibitively high.

If a well-to-do land-reform beneficiary in a well-irrigated area desires to minimize his own farm-work, including labour supervision, the most efficient way is to have a tenancy contract with a worker, leaving all the tasks to him for a fixed or a share rent. In fact, some new leaseholders have opted to have subtenancy contracts with landless workers (Hayami and Kikuchi 1982, chs. 5 and 6). However, the subtenancy arrangements are highly hazardous because if the lessee should appeal to the Agrarian Reform Office, the lessor's tenancy title would be forfeited and transferred to the lessee if the latter proved himself to be an actual 'tiller of the land'. Therefore, subtenancy arrangements are bound to be limited to a narrow circle of relatives and close friends.

The new permanent labour of the semi-attached type can be considered an institutional innovation to fill the demand of well-to-do land-reform beneficiaries in irrigated areas, who want to withdraw from farm-work while keeping their title to the land. Such labour plays a role similar to a tenancy contract, while it can be claimed as a labour-employment contract.

Transfer of land-cultivation rights to urban residents

Permanent labour as a substitute for tenancy contracts has also been used by big landlords with sizeable tracts of land under their direct administration because non-tenanted lands were exempt from land-reform programmes.[3] More recently it has been used by middlemen and money-lenders who have acquired the land-cultivation rights from land-reform beneficiaries through an illegal pawning practice while trading in rice, fertilizer, and other farm inputs. Under this practice a land-reform beneficiary, in return for a certain amount of credit, surrenders his land to his lender's cultivation until he repays the debt. This practice has emerged partly because CLT and leasehold title are not transferable except to legitimate heirs so that the amortizing owner and leaseholder cannot use them as collateral for institutional loans.

Since this practice is illegal, a survey like the Central Luzon Loop Survey based on a standard questionnaire to cover a wide area tends to fail in capturing the incidence of this credit–land transfer arrangement, but the intensive survey on the Nueva Ecija village reveals that this arrangement is not uncommon (Table 9.1).[4] Yet the intensive village survey data shown in Table 9.1 also tend to underestimate the incidence of land-pawning because these data pertain to the plots cultivated by farmers living in this village. In fact, land-reform beneficiaries pawn out their cultivation rights not only to other farmers in the same village but also, more frequently, to money-lenders living outside the village (Table 9.2); these lenders are mostly middlemen/money-lenders living in nearby local towns.

Those middlemen/money-lenders as well as many landlords who live in the towns need agents in villages to take care of their farms. Under the land-reform laws ruling out the use of tenancy contracts, they have had no other choice but to enter permanent-labour contracts, even if the latter are an efficient substitute for the former. As will be shown, as many as 73 per cent of permanent labourers were employed by non-village residents, 78 per cent by pawnees, and 39 per cent by traders/money-lenders in the Nueva Ecija village in 1989.

If there were no land-reform regulations, these urban dwellers, for whom the cost of monitoring labourers' work is high, would have opted for tenancy rather than permanent-labour contracts. This hypothesis is consistent with earlier experience in Central Luzon; during the Spanish period Chinese (and mestiso) traders who acquired land-cultivation rights through a pawning arrangement similar to that practised today continued to manage their farms

TABLE 9.2. *Matrix of land transfer through pawning of CLT and leasehold titles in the Nueva Ecija village, 1987 (no. of transfers)[a]*

Farm-size class (ha.)	Pawn-in						
	Villager (farm size class: ha.)				5 and above	Non-villager	Total
	Below 0.9	1–1.9	2–2.9	3–4.9			
Pawn-out							
Below 0.9		1 (0.2)					1 (0.2)
1–1.9				2 (1.5)	2 (2.0)	1 (0.75)	5 (4.25)
2–2.9				4 (3.9)	1 (1.0)	4 (5.7)	9 (10.6)
3–4.9	1 (1.0)	1 (1.0)	1 (1.0)	1 (1.0)	1 (1.5)	5 (7.0)	9 (11.5)
5 and above				1 (1.0)	1 (4.0)	1 (1.0)	3 (6.0)
TOTAL	1 (1.0)	2 (1.2)		8 (7.4)	5 (8.5)	11 (14.45)	27 (32.55)

[a] Hectares of transferred areas are shown in parentheses.

Source: Based on the baseline survey for the Nueva Ecija Village Survey.

under share-tenancy contracts with the natives who pawned out their lands (McLennan 1969; Hayami and Kikuchi 1982, ch. 4).

'Indianization' of the Philippine rice bowl

The process of diffusion of new permanent-labour contracts observed for Central Luzon suggests a new agrarian structure toward which the irrigated rice sector in the Philippines is now moving.

Before land reform and new technology, the rice sector of the Philippines was essentially divided between wealthy landlords and poor share-tenants/agricultural labourers. In general, share-tenants and agricultural labourers made a continuous social spectrum and were linked by the agricultural ladder. With land reform, not only land assets but also incomes from land were redistributed in favour of tenants through regulations on land rent and form of tenancy. Since then the incomes of land-reform beneficiaries have increased rapidly, corresponding to the development and diffusion of new rice technology. These large benefits have been limited to ex-share tenants and have almost completely bypassed landless agricultural labourers (Otsuka *et al*. 1991).

Moreover, the agricultural ladder has been closed for agricultural labourers because neither land-reform beneficiaries and landlords who still administer sizeable areas directly, nor the urban rich who acquired cultivation rights, have any incentive to rent out their lands under present land-reform regulations. While the possibility of climbing up the agricultural ladder has been closed, the possibility of dropping down has been left open. It is commonly observed that small leaseholders and amortizing owners are forced to sell or pawn out their titles for urgent need of cash and thereby drop down to the landless-labour class.

Thus the irrigated rice sector in the Philippines has been divided between non-working farmers/semi-landlords and landless agricultural labourers. While the increasing disutility of labour and the increasing preference for non-farm economic activities have been encouraging wealthy farmers to reduce their family labour input in farm-work, the option to rent out their land to landless labourers has been closed under the land-reform regulations. At best, landless labourers have been forced to get a meagre living from permanent-labour contracts, and at worst from unstable, casual farm-work.

The emerging agrarian structure in the Philippine rice sector resembles that of India in which farmers in upper castes do not work themselves but only supervise the work of labourers in lower castes, with no agricultural ladder bridging them. This 'Indianization' of the Philippine rice bowl in recent years suggests strongly the hypothesis that the new permanent-labour contract has emerged as an inefficient substitute for the land-tenancy contracts under the limitations of land-reform laws.

9.4 Tests on Relative Income and Efficiency of Permanent Labour

We now move to empirical testing of the hypothesis that, being an imperfect substitute, the permanent-labour contract is less efficient than tenancy as well as owner-farming. Specifically we examine the following hypotheses based on the data of the Nueva Ecija

Village Survey (the site indicated in Fig. 9.1): (*a*) permanent labourers are better off than casual labourers because employers of permanent labourers offer them utility higher than their reservation utility to elicit loyal work-effort, as argued in the theories of permanent-labour contract (Section 4.2); (*b*) permanent labour is not a perfect substitute for family labour of tenant- and owner-farmers because of insufficient work-incentives; and (*c*) as a result, production and profit per hectare are lower on the farms based on family labour than on the farms based on permanent labour.

Characteristics of sample farmers, labourers, and employers

Our stratified random sample survey in 1989 covered 62 'pure' farmer households (leaseholders, CLT-holders, and owner-farmers), 9 households of 'farmer-cum-permanent labourer' (farmers who work a part of their time as permanent labourers on others' farms besides working on their own farms), and 23 landless permanent-labourer and 25 casual-labourer households. In the statistical analysis in this study we classify 71 farms based on the family labour of 62 'pure' farmers and 9 farmer-cum-permanent labourers as the 'farms of farmers' family-labour operation'. The remaining 36 farms operated under permanent-labour contracts, by which either 9 farmer-cum-permanent labourers or 23 'pure' permanent labourers are employed, are classified as the 'farms of permanent-labour operation'. Four among these labourers took care of two farms each. With no exception, only one permanent labourer is employed in each farm.

According to Table 9.3, the 'pure' farmers were the eldest and most experienced in rice-farming, whereas the 'pure' permanent labourers were the youngest and least experienced. Reflecting the cohort effect, however, the 'pure' farmers were not the most educated in terms of schooling. There was no significant difference in age, experience, and schooling between the permanent and casual labourers, which suggests that these two types of labourers were relatively similar.

All farmers and landless labourers in our sample were married and heads of households. The families were generally nuclear, consisting of one married couple and their children. The average size of the farms of permanent-labour operation was larger than that of farmers' family-labour operation; this reflects the fact

TABLE 9.3. *Socio-economic characteristics of sample farmers, permanent labourers, and casual labourers in the Nueva Ecija village, 1989 dry season*[a]

	Farmer	Farmer-cum permanent labourer	Permanent labourer	Casual labourer
Sample size	62	9	23	25
Age	50.4	46.9	37.2	43.6
	(12.4)	(15.9)	(11.3)	(17.2)
Years of farming	37.0	34.6	23.8	29.7
experience	(14.2)	(18.2)	(13.2)	(18.7)
Years of schooling	6.4	5.3	6.4	6.9
	(3.3)	(2.8)	(3.0)	(3.0)
Family size	6.9	6.6	5.8	4.9
	(2.1)	(2.2)	(2.5)	(1.7)
Operational farm	1.95	2.19	3.27	—
size (ha.)	(1.34)	(0.57)	(2.27)	

[a] Numbers in parentheses are standard deviations.

that family members worked together on the farm under the permanent-labour contract, while a larger portion of farmers' family members engaged in non-farm work and schooling. Partly reflecting the difference in employment opportunity, the family size of permanent-labourer households is larger than that of casual-labourer households.

About three-quarters of employers of permanent labourers were non-residents and pawnees (Table 9.4, col. 3). Nearly 40 per cent of them were traders/money-lenders while heirs of hacienda-owners were merely 20 per cent. This suggests that a new class structure has been emerging through pawning operations in rural areas of Central Luzon as a result of rapid technological changes in rice-farming and land-reform regulations on tenancy contracts.

There were three non-resident employers of permanent labourers managing five sample farms who engaged in farm management as a full-time job. There were also five sample farms operated by permanent labour under the supervision of professional managers employed by the farm-owners. Unlike other employers, who typically manage only a few farms and visit their farms occasionally, these six professional employer/managers commute to the village almost every day to manage on average as many as seven farms.

TABLE 9.4. *Socio-economic characteristics of employers of permanent labourers in the Nueva Ecija village, 1989 dry season*[a]

	Ordinary management (1)	Professional management (2)	Total (3)
Sample size	26	10	36
Ratio of non-resident (%)	62	100	73
	(47.1)	(0)	(43.7)
Ratio of traders/money-lenders (%)	15	100	39
	(37.3)	(0)	(49.2)
Ratio of pawnees (%)	69	100	78
	(47.1)	(0)	(42.4)
Ratio of heirs of hacienda-owners (%)	4	70	22
	(20.0)	(45.8)	(42.4)
Years of schooling[b]	10.3	13.7	11.2
	(4.3)	(0.9)	(5.1)
Number of farms managed per manager	1.9	6.8	3.1
	(1.3)	(2.3)	(2.7)

[a] Numbers in parentheses are standard deviations. Employers hiring more than one permanent labourer are counted more than once.
[b] Pertaining to those who actually engage in management of the farms.

(They manage also the farms not included in our sample.) They had all completed college education and hence were more educated than ordinary employers (see Table 9.4). Moreover, two of them had a bachelor degree in agricultural sciences and they often exchanged information on farming practices with each other. In view of their superior farm-management abilities, we distinguish those ten farms under 'professional' management from the other twenty-six farms under 'ordinary' management.

Payment to permanent labourers was 10 per cent of gross output in nearly 85 per cent of cases and fixed in the rest. We do not distinguish them in this study because we did not find any major difference in the characteristics of labourers, the scope of assigned jobs, the characteristics of their employers, or their production performance.

Relative incomes of permanent versus casual labourers

First we examine the hypothesis that *the permanent labourers are better off than ordinary casual labourers*. For this purpose we compared the labour income and workdays of twenty-three permanent and twenty-five casual labourers during a period of five months from January to May 1989 (Table 9.5).

TABLE 9.5. *Labour income and workdays of permanent and casual labourers in the Nueva Ecija village by source, 1989 dry season*

	Permanent labourer	Casual labourer	Student t-statistics
Sample size	23	25	—
Labour income (₱/season):			
Rice production	7,594 (6,724)[a]	802	7.33**
Non-rice production	1,138	3,533	−2.27*
TOTAL	8,732	4,335	3.31**
Workdays:			
Rice production	71.2 (56.4)[a]	20.8	4.81**
Non-rice production	23.0	79.4	−3.02**
TOTAL	94.2	100.2	−0.28

[a] Numbers in parentheses refer to employment under permanent-labour contract.
* Difference is significant at the 5% level.
** Difference is significant at the 1% level.

The major sources of income of these landless labourers were permanent and casual labour in rice production and unskilled off-farm jobs, such as carpentry and tricycle operation. We excluded nine farmer-cum-permanent labourers from the comparison because their income consisted not only of returns from labour but also of returns from land. The total labour income of permanent labourers (₱8,732) is found to be twice as large as that of casual labourers (₱4,335), and the difference is highly significant. In terms of total workdays, however, there is no significant difference between permanent and casual labourers.[5] The significant difference in total income arises mainly because the estimated earnings per day under the permanent-labour contract is far greater than the daily wage-rate for casual labour.[6] Thus it seems clear that permanent labourers are better off than casual labourers, owing to the lucrative terms and conditions of the permanent-labour contract.

This difference cannot be attributed to the difference in labour quality between permanent and casual labourers because there is no significant difference in average age, schooling, and farming experience. Furthermore, there is no significant difference in average labour income per day from casual employment between permanent and casual labourers, which suggests the similarity of labour quality between the two categories of labourers. Thus the significantly higher income offered to permanent labourers can be considered a device to elicit loyal work-efforts from them.

Relative efficiencies of family labour versus permanent labour

We now examine the second hypothesis that *permanent labour is an imperfect substitute for the family labour of tenant- and owner-farmers*.

If permanent labourers work less hard or less conscientiously than tenant- and owner-farmers, it is expected that the labour input per hectare by permanent labourers and employers themselves in terms of workdays on the farms of permanent-labour operation are lower than on the farms of farmers'-family-labour operation. While workdays are relatively easy to measure for employers of permanent labourers, the intensity of work-effort, particularly for tasks requiring care and judgement, is difficult to monitor without close supervision. We expect that employers spend considerable time in supervising their permanent labourers, aside from the supervision of casual labourers.

According to Table 9.6, the labour input of farmers and their family members in farms without permanent labour was 33.1 days per hectare per season, whereas labour input of employers and hired managers of permanent labours was only 6.0 days in the case of ordinary management and 5.2 days in the case of professional management. The low level of labour input in the latter was to a large extent compensated for by the input of permanent labourers. Yet the farmer's labour input in the farms of family-labour operation was significantly larger than the sum of the employer's and permanent labourer's labour input in the farms of permanent-labour

TABLE 9.6. *Comparison of labour inputs per hectare between the farms of farmers' family-labour operation and of permanent-labour operation in the Nueva Ecija village, 1989 dry season*

	Family-labour operation	Permanent-labour operation		Student t-statistics		
		Ordinary management	Professional management	(1) vs. (2)	(1) vs. (3)	
	(1)	(2)	(3)			
Sample size	71	26	10	—	—	
Operational farm size (ha.)	1.8	2.5	2.3	—	—	
(Man-days/ha./season)						
Farmer/employer[a] (A)	33.1	6.0	5.2	9.16**	9.12**	
Permanent labour (B)	0	21.0	17.6	—	—	
Subtotal (A+B)	33.1	27.0	22.7	1.79*	2.66**	
Casual labour	37.7	47.0	56.6	−2.24*	−3.20**	
TOTAL	70.8	74.0	79.3	−0.58	−1.02	
Labour supervision	2.2	5.6 (0.7)[b]	5.9 (0.8)[b]	−6.33**	−5.02**	
Land preparation and crop care[c]	21.9	20.9	17.9	0.44	1.18	
Transplanting, harvesting, and threshing	46.7	47.5	55.7	−0.20	−1.46	

[a] Includes family members of farmer households, employers, and hired managers of permanent labourers.
[b] Supervision of casual labourers by permanent labourers.
[c] Includes irrigation water control, fertilizer and chemical applications, and land preparation.
* Difference is significant at the 5% level.
** Difference is significant at the 1% level.

operation. This gap was filled by casual labour. These observations suggest that not only permanent labour but also casual labour was substituted for family labour.

Although the total number of workdays was smaller in the farms of family than in the farms of permanent-labour operation, the difference is not significant. There is, however, a highly significant difference in labour-supervision time. In fact, almost all the labour input of employers and managers of permanent labourers was that of supervising permanent as well as casual labourers. The farmers of family operation who employed casual labourers only occasionally spent much less time. These observations clearly support the hypothesis that direct supervision is needed for monitoring permanent labour because of insufficient work-incentives.

There is little scope for substitution of capital for labour in the Nueva Ecija village because tractors and threshers were used on all farms. Direct seeding, which was adopted by 60 per cent of farms, was the chief way of reducing labour input as it saves the work of transplanting. Since transplanting is easy to monitor, casual labour is mostly used so that the adoption of direct seeding is largely unrelated to the employment of permanent labour. Indeed, this method was practised only in fields with better water control and there was no significant difference in its adoption rate between the farms with and without permanent labour. The labour input of harvesting and threshing in farms under professional management was substantially higher than in other farms because of the significantly higher physical yield of farms under professional management (7.0 tonnes/ha.) compared with farms of family operation (6.3 tonnes/ha.), and farms of permanent labour operation under ordinary management (5.5 tonnes/ha.).

Overall the results of labour-use analysis support the hypothesis that permanent labour is an imperfect substitute for the family labour of tenant- and owner-farmers.

Comparisons in output and profit

Lastly we examine the hypothesis that *production and profit are lower on farms with than farms without permanent labour*.

Table 9.7 shows average output values, production costs, and residual profits per hectare in three types of farms. We consider the value of output rather than the physical yield because there

TABLE 9.7. *Comparison of average output value, production costs, and residual profit per hectare between farms with and without permanent labour in the Nueva Ecija village, 1989 dry season (₱1,000/ha.)*

	Family-labour operation (1)	Permanent-labour operation		Student t-statistics	
		Ordinary management (2)	Professional management (3)	(1) vs. (2)	(1) vs. (3)
Gross value of production (A)	25.5	22.4	26.8	3.46**	-0.94
Current input costs[a] (B)	3.1	3.0	3.3	0.72	-0.67
Capital costs[b] (C)	1.8	1.7	1.8	0.79	0.07
Labour costs[c] (D)	4.1	4.9	5.5	-3.47**	-4.17**
Supervision costs[d] (E)	0.2	0.6	1.1	-5.34**	-4.00**
Residual profit (A-B-C-D-E)	16.4	12.2	15.2	5.32**	0.97

[a] Total costs of chemical fertilizer, other chemical inputs, fuel, and seeds.
[b] The sum of actual and imputed rental costs of draught animals, tractors, and threshers.
[c] The sum of hired-labour costs, including payments to permanent labourers, and imputed family labour costs.
[d] Man-days of supervision time multiplied by the estimated time costs per day.
* Difference is significant at the 5% level.
** Difference is significant at the 1% level.

were marked differences in prices of paddy among varieties grown in the Nueva Ecija village. The production in the 1989 dry season was particularly good because of the absence of damage to crops from insects, disease, and other natural calamities. The absence of variations in output due to these uncertain factors increases the likelihood that the inefficiency of the permanent-labour contract, if any, would be detected statistically.

The average value of output on farms of family operation (col. 1) is significantly higher than on farms of permanent-labour operation under ordinary management (col. 2). This observation is consistent with our hypothesis that permanent labourers shirk because of the weak work-incentives. The average value of output on farms under professional management (col. 3), however, is comparable with that on farms of family operation. This may reflect the superior ability of these professional managers which more than makes up for the disincentive effects of the permanent-labour contract.

The value of output depends not only on the work-effort of permanent labourers and the quality of management inputs but also on the application of other inputs and supervision of permanent labourers by their employers. We therefore compared the costs of current, capital and labour inputs among three farm categories. Further, we estimated the residual profits by subtracting these costs and the cost of supervision from the output value. The costs of current inputs include costs of chemical fertilizer, other chemicals, fuel, and seeds. Costs of capital services include actual rental costs of tractors, draught animals, and threshers as well as the imputed costs of owned capital using market rental rates. Similarly, labour costs consist of payments to casual and permanent labourers and the imputed costs of family labour, estimated by using the prevailing market wage-rates. For the evaluation of the time-cost of supervision per day, we applied ₱80 for resident farmers, ₱120 for non-resident, ordinary managers of farms with permanent labour, and ₱200 for the professional managers. The imputed cost of ₱80 corresponds to the average daily earnings of harvesters under output-sharing contracts, which is the highest hired-labour wage in rice production in the Nueva Ecija village. The estimated time-costs of managers of permanent labourers are obtained from interviews with professional as well as ordinary managers.[7] The residual profit is considered a rough measure of farming efficiency.

The costs of current inputs and capital inputs were largely the same across farm categories. As was shown in Table 9.6, the total workdays per hectare were no different, particularly if we exclude supervision time. The quality of land is also considered largely the same because all paddy fields are located in flat topography and fully irrigated by gravity irrigation in this village. Nevertheless, the output value in the farms of permanent-labour operation under ordinary management was significantly lower than on farms of family operation. These observations strongly suggest that the work-effort of permanent labourers was significantly lower than that of tenant- and owner-farmers. If so, the fact that the output value on farms under professional management was much larger than on farms under ordinary management must be due to the difference in managerial ability between professional and ordinary managers.

Labour costs were higher on farms of permanent-labour operation largely because of the larger wage payments to permanent labourers. There were also significant differences in supervision costs. As a result, the residual profit on farms of permanent-labour operation under ordinary management amounts to only 74 per cent of the profit on farms of family operation. Such a result is consistent with our hypothesis that, being an imperfect substitute for tenancy, farming under the permanent-labour contract is less efficient than under tenant- and owner-farming because of insufficient work-incentives under the fixed-wage contract and of the inherent difficulty of supervising permanent labourers in spatially dispersed and ecologically diverse agricultural production environments.[8]

On the other hand, both the output and the residual profit of the farms employing permanent labour under professional management were not significantly different from those of the family-operated farms. This finding indicates that inefficiency arising from the substitution of permanent labour for family labour could be corrected by the use of superior management. However, this correction could only be achieved at the expense of human resources that commands a very high opportunity cost. Relative inefficiency is also likely to be found on farms of permanent-labour operation under professional management if the high cost of management is properly accounted for, even though such calculation is not possible from the available data.

9.5 Conclusion

A major thrust of this book is to argue that agrarian organizations are better understood by treating labour-employment and land-tenancy contracts as alternatives along a spectrum of contract choice rather than treating them separately. From this perspective, we advance the hypothesis from our theoretical analysis (Chapter 4) that, in peasant economies characterized by the absence of scale economies, permanent-labour contracts are chosen where institutional constraints preclude land tenancy from the scope of contract choice. In this chapter the hypothesis is tested by a recent study of the rice sector in the Philippines where the new form of permanent labour has been spreading rapidly. The results strongly support the hypothesis that permanent labour has been employed as an inefficient substitute for tenancy contracts because of legal restrictions on the choice of tenancy.

These findings are considered evidence to support a more general postulate put forth in Chapter 6 that artificial limitations on contract choice in agrarian economies will lead to inefficient alternatives as substitutes. It is not a mere coincidence that we find inefficiency in resource allocation under share tenancy in some parts of India and Bangladesh, where land-to-the-tiller legislation has discouraged landlords from offering fixed-rent leasehold tenancies for fear of possible confiscation of leased land, and under the permanent-labour contract in the Philippines, where tenancy regulations have precluded the option of subtenancy.

Our analysis in this chapter also indicates that the artificial limitation of contract choice in agrarian economies results not only in inefficient resource allocation but also in major disadvantages to the landless poor by closing the agricultural ladder for them to ascend from wage-workers to share-tenants and, further, to leasehold-tenants.

Notes to Chapter 9

1. After completion of amortization payments, the CLT-holder is given the Emancipation Patent (EP) that entails landownership transferable only to his legitimate heirs.
2. The land to be transferred from landlords to CLT-holders was valued at 2.5 times the average output in three normal crop years immediately

preceding 1972. As Mangahas (1985) has demonstrated, the annual amortization fee, the discounted sum of which at 6% annually for 15 years amounts to the land value, is approximately the same as the leasehold rent prescribed by law.
3. A significant area is reported to have been added to the land under landlords' direct administration through illegal eviction of tenants (Otsuka 1991).
4. Theoretically, the government, now the formal owner of the CLT lands, can forfeit the rights of tenants who engage in illegal practice, and transfer the rights to appropriate cultivators. In practice, it is not easy to prepare evidence that can stand up to lawsuits because the pawning contract is usually disguised as a simple credit contract. Moreover, the legal process must go through from mediation by village councils to municipal courts and then to higher courts. Since the pawning contract has the informal sanction of the village community and is usually approved by a village captain or councilman with formal signature, it is extremely difficult, if not impossible, to impose a penalty. Because the administrative burden of taking legal action against the pawning practice is likely to be prohibitive, local agrarian-reform officers tend to close their eyes to the practice.
5. Note that although relatively unimportant, workdays of family members of permanent labourers are included in the workdays under the permanent-labour contract.
6. In India Binswanger *et al.* (1984) found evidence consistent with our finding, whereas earlier studies by Ghose (1980), Bardhan (1983), and Basant (1984) found that the estimated earnings per day were lower than the casual daily wage, even though the permanent labourer's total income was larger than the casual labourer's.
7. Our estimates are based primarily on interviews with five managers who were hired by employers of permanent labourers. The payments to them, however, were not fixed and took various forms including benefits in kind. The estimated time-costs therefore are subject to errors. However, the qualitative results of our analysis remain unchanged even if we apply the uniform rate of time-cost because of the large difference in supervision time.
8. We also performed the regression analysis of the gross value of production and the residual profit using farm size, a permanent-labour contract dummy, a professional management dummy, and variables pertaining to the quality of labour and land inputs as independent variables. Since the estimated coefficients of input-quality variables are not significant, there is no inconsistency between the regression analysis and the analysis of the difference in means.

10

Towards a General Theory of Agrarian Contracts

IN search for a general theory of contract choice in agrarian economies, we have reviewed critically existing theories and tested their relevance in the light of evidence collected from a global survey of past empirical studies as well as from our case-studies. Existing theories have demonstrated markedly divergent explanations of the causes and the effects of specific contract choice, especially with respect to share-cropping tenancy and permanent-labour employment. Much confusion has stemmed from partial treatments of contractual choice which unduly limit both the optimizing behaviour of contracting parties and the options of forms of contract.

10.1 Theoretical Conclusions

We have attempted to identify the sources of this confusion by clarifying the limitations of existing theories in the light of the more general theory of agent–principal relations. Major conclusions drawn from our theoretical analysis (Chapters 2 to 5) may be summarized in the following four postulates:

1. *Where stipulated contract terms are difficult for landowning principals to enforce and landless farm-workers are risk-neutral, fixed-rent contracts will be chosen, whereas share contracts will be chosen when workers are risk-averse.* Because the enforcement of a worker's effort alone is the inherent contract problem in the absence of risk-aversion, the landlord prefers the fixed-rent contract because it does not distort work-incentives. This holds true regardless of the length of contract. Thus not only the Marshallian theories of share tenancy but also Eswaran and Kotwal's (1985*a*) permanent-labour contract model precludes the option of the fixed-rent contract in order to justify the existence of less efficient alternative contracts under certainty. If risk-aversion is present, an

increase in the worker's output-sharing rate enhances his work-incentives only at the cost of increasing his income risk. Given such a trade-off, the share contract is likely to be chosen in equilibrium. This theory provides the most consistent explanation for the existence of the share contract. Furthermore, it rules out the possibility of the permanent-labour contract existing under the normal trade-off relation between risk and work-effort; it can be chosen only in a world of perfect contract enforceability so long as tenancy contracts are available options. The effort–management trade-off model of Eswaran and Kotwal (1985b) and the effort–land-abuse trade-off model of Murrell (1984) and Datta *et al.* (1986) also rationalize the choice of a share contract without assuming risk-aversion but only within a short-term context. If the contract is long-term and reputation plays a significant role in contract enforcement, as is commonly the case in agrarian economies, undersupply of management input and abuse of land tend to be deferred. Thus it seems difficult to justify the prevalence of share contract based on such trade-off relations alone, even though they may add to the incentive for the choice of share tenancy.

2. *Where the farm-worker's effort is perfectly enforceable, optimum contracts are, in general, indeterminate.* If the worker is risk-neutral and contract enforcement is costless, as in Cheung's (1969) share-tenancy model and in Bardhan's (1979b) permanent-labour contract model, there is no contract problem. In fact, all forms of contract become equally efficient with identical distribution of expected output in equilibrium. If both the worker and the landlord are risk-averse under the enforceable contract, the share contract will be chosen in order to share production risks. However, the share contract can be dispensed with because a suitable linear combination of a fixed-rent and a fixed-wage contract can mimic the desired share contract.

3. *By increasing the options of punishment and reward, long-term and interlinked contracts provide added work-incentives to the farm-worker.* In an agrarian community where social interaction is intense and various transactions are interlinked, reputation is likely to play a major role in contract enforcement in a long-term context. Moral hazard or breaches of contract are likely to be detected in the long run when parties engage in regular transactions and, once detected, the reputation of the defaulting party is harmed in the community thus reducing future contract opportunities and welfare.

When the expected loss of future utility is sufficiently great, opportunistic behaviour becomes unlikely and the worker's effort comes closer to the first-best outcome in which his effort can be monitored.

4. *Major puzzles, such as the prevalence of the 50:50 sharing ratio in tenancy and the absence of fixed payments in share tenancy, the equality of output and cost-sharing rates, and the low interest rate charged on credit provided by the landlord to his tenant/labourer, cannot be understood without carefully considering the interlinking of contracts.* The absence of fixed payments is easily resolved because credit repayment and shared cost under the cost-sharing arrangement can be considered substitutes for the fixed payment. Existing models of interlinked contracts do not go very far in providing an explanation for the other empirical regularities. However, our review suggests that consideration of appropriate discounting of contracting parties' incomes at the time of harvest with the use of interest rates charged on the interlinked credit may be necessary for an understanding of these regularities. However, sociological factors are needed for an adequate explanation of rural people's preference for the 50:50 share rate (augmented with complicated adjustments through interlinked transactions) to simpler adjustments in the share rate itself.

10.2 Empirical Conclusions

The theoretical conclusions enumerated above leave a number of questions to be resolved. The major questions for our empirical analysis are: Does share tenancy result in significantly inefficient resource allocation relative to owner-farming and fixed-rent tenancy? Is land rent higher under share tenancy than under fixed-rent tenancy? What are the conditions in which permanent-labour contracts emerge? Conclusions drawn from both the global survey and our case-studies are as follows:

1. *Significant inefficiency in share tenancy is not commonly found in areas where both share and fixed-rent contracts are available options: inefficiency tends to arise where the contract choice is institutionally restricted.* This empirical regularity found from the global survey is largely consistent with the general hypothesis that, when choices are available to them, rural people in developing economies make efficient choices from a wide spectrum of agrarian

contracts. This finding does not mean that contract enforcement is without cost. Rather, the observed efficiency of share tenancy compared with fixed-rent tenancy reflects the tendency that the former is chosen by landlords endowed with the ability to monitor tenants' work-effort at low opportunity cost while the latter is chosen by those who lack such ability; it also reflects the tendency that the former is chosen in relatively closed communities where social interactions are intense and personal relations are enduring and multi-stranded. These tendencies are clearly indicated in our case-studies (Chapters 7 and 8).

2. *Rent is significantly higher under the share than the fixed-rent contract.* This general tendency observed from both the global survey and our case-studies confirms that agrarian economies are characterized by high risk due to weather hazards as well as to small market size, in addition to the fact that landless farm-workers are usually risk-averse.

3. *The fixed-wage permanent-labour contract is commonly found in agrarian economies where land tenancy is illegal, while the tenancy contract predominates in the absence of such legislative constraints.* This seems to reflect the general difficulty of landowning principals in enforcing contracts with landless agents so that low work-incentive contracts such as the fixed-wage permanent-labour contract is more common when land-tenancy contracts are unavailable. In other words, the transaction costs of permanent-employment contracts are usually so high relative to those of land tenancy that small tenant-farms are more common than large-scale farm firms in agrarian economies in the absence of restrictions on the choice of tenancy contracts. Relative inefficiency in resource allocation under permanent-labour contracts as compared with tenancy contracts is supported by our case-study for the Philippine rice area under land-reform programmes (Chapter 9).

4. *In general, rural people in developing economies make efficient choices from a wide spectrum of contracts, ranging from casual-labour employment to long-term fixed-rent tenancy, with due consideration of their own resource endowments and the external conditions surrounding them.*

10.3 Tasks Ahead

A number of tasks have to be completed before a general theory of contract choice in agrarian economies can be formulated.

In theory, successful modelling has assumed the homogeneity of human capital endowments of both landed principals and landless agents. This assumption has been instrumental in producing models to explain effectively the differences in the general pattern of contract choice across regions and across historical epochs with different institutional and technological environments (e.g. with or without land-reform programmes and with or without modern high-yielding varieties). This assumption, however, ruled out the possibility of building models to explain different contract choices by agents and principals with different human capital endowments within the same institutional and technological environment. The self-selection model intended to explain different contract choices based on heterogeneity in agent's managerial ability has not been successful so far. The classic theory of the agricultural ladder has remained a verbal hypothesis. In order to build a general model of land and labour contracts in agrarian economies, major efforts will be necessary to incorporate the heterogeneity of both principal and agent in order to explain different contract choices both within the same institutional and technological environment and across the different environments. Although progress in this direction has already been made (e.g. Eswaran and Kotwal 1986; Bell and Zusman 1989), much research still remains to be done.

It must also be pointed out that existing studies do not pay sufficient attention to the operation of market for land. The role of the land market is highly relevant in the study of tenancy, because land may be sold to the tenant if tenant cultivation is less efficient than owner cultivation. The analysis of the land market, however, has been largely ignored in the literature on agrarian economies.

Models of contract choice in agrarian economies have assumed the absence of scale economies. In this framework the fundamental question of the dominance of small-scale family farms (either owner- or tenant-operated) can be explained by higher enforcement costs of hired wage-labourers than tenants in a typical agricultural environment; this disadvantage of labour contracts is considered to increase with farm size. But large-scale farms based on hired labour do exist and often coexist with small-scale farms, as typified by the latifundio–minifundio complex in Latin America. In order to gain a more satisfactory understanding of the dominance of family farms, we have to explore the conditions under which the advantages of large operational units emerge and outweigh the disadvantages of labour contracts. If the operational scale advantages are associated with the advantages of large land ownership, such as better access to technological information and capital markets, large farm firms or plantations may dominate family farms, despite high transaction costs of labour-employment relative to land-tenancy contracts. In order to build a truly general theory of agrarian organization capable of explaining large variations in farm size and tenure distribution across countries and regions as illustrated in Tables 1.1 and 1.2, we have to broaden our perspective on contract choice by incorporating the causes of differential access among rural people to technology, and factor and product markets. Attempts in this direction have so far been partial without considering the full range of contract choice (e.g. Feder 1985; Eswaran and Kotwal 1986).

Theoretical development in this direction must be paralleled with empirical investigations. Many of the past empirical studies simply compared resource allocation between share tenancy and owner-farming or leasehold tenancy exclusively to test the validity of Marshallian hypothesis without a due consideration of contract choice in the light of institutional constraints and heterogeneity in the ability of landowning principals to monitor the work-effort of landless agents. An integrated analysis of the determinants of contract choice and relative efficiency of alternative contracts is clearly called for.

At present, a major bottleneck for theoretical advancement is the lack of adequate information on interlinked contracts. A better understanding of the borrowing behaviour of the tenant/labourer in regular credit markets and from the landlord/employer is badly needed. Also imperative is the analysis of the collateral value of the farm-worker's expected income in the interlinked credit transactions in relation to repayment patterns of credit over time. The major efforts to identify the prevailing modes of interlinked exchanges with respect to repayment, interest rate, and collateral based on long-term survey of rural household economies is critically needed. Such economic investigation must be paralleled by an anthropological and sociological approach to social norms in rural communities for building the general theory of contract choice in agrarian economies.

It must be pointed out that the theories of agrarian contract, of which we have tried to develop a synthesis in this volume, are not only relevant to agrarian economies but also to many sectors of urban economies, such as commerce, personal and professional services, characterized by the narrow scope of technological scale economies and the difficulty of monitoring the work-effort of hired workers. In the industrial sector, too, as the work required in the modern world has shifted from that based on muscles to that based on brain, it has increasingly become difficult to enforce workers by a hierarchical command system. It has become necessary to design forms of contract that incorporate incentives to enhance unobservable work-effort. One possible direction is to establish relations of a community type within a firm. A typical example along this line is the Japanese management system. In this system employment is lifelong with no explicit contract but both administration and employees are assumed to follow the customary rules of the company; a boss is supposed to develop a patron–client relationship with workers under him so that a section or a division or even a whole company simulates a family or a village. Such a system, which was once regarded as pre-modern, feudalistic, and hence inefficient, has been recently considered to underlie the high efficiency of Japanese industries (Hirschmeir and Yui 1981; Aoki 1984; Leibenstein 1987).

It would, therefore, not be an unreasonable expectation that the efforts to build a general theory of agrarian contracts will also push frontiers in the theory and the design of industrial organization in advanced economies.

References

Abdullah, Abu (1976), 'Land Reform and Agrarian Changes in Bangladesh', *Bangladesh Devel. Stud.*, 4(1), 67–114.

Adams, Dale W., and Rask, Norman (1968), 'Economics of Cost-Share Leases in Less-Developed Countries', *Amer. J. Agr. Econ.*, 50(4), 935–42.

Ahmad, Mushtaq (1974), 'Farm Efficiency under Owner Cultivation and Share Tenancy', *Pakistan Econ. Soc. Rev.*, 12(2), 132–43.

Akerlof, George A. (1976), 'The Economics of Caste and of the Rat Race and Other Woeful Tales', *Quart. J. Econ*, 90(4), 599–617.

Alchian, Armen A., and Demsetz, Harold (1972), 'Production, Information Costs, and Economic Organization', *Amer. Econ. Rev.*, 62(5), 777–95.

Allen, Franklin (1982), 'On Share Contracts and Screening', *Bell J. Econ.*, 13(2), 541–47.

—— (1984), 'Mixed Wage and Rent Contracts as Reinterpretation of Share Contracts', *J. Devel. Econ.*, 16(3), 313–17.

—— (1985), 'On the Fixed Nature of Sharecropping Contracts', *Econ. J.*, 95(377), 30–48.

Alston, Lee J. (1981), 'Tenure Choice in Southern Agriculture, 1930–1960', *Exploration. Econ. Hist.*, 18(3), 211–32.

—— Datta, Samar K., and Nugent, Jeffrey B. (1984), 'Tenancy Choice in a Competitive Framework with Transactions Costs', *J. Polit. Econ.*, 92(6), 1121–33.

—— and Ferrie, Joseph P. (1985), 'Labor Costs, Paternalism, and Loyalty in Southern Agriculture: A Constraint on the Growth of the Welfare State', *J. Econ. Hist.*, 45(1), 95–117.

—— and Higgs, Robert (1982), 'Contractual Mix in Southern Agriculture since the Civil War: Facts, Hypotheses and Tests', *J. Econ. Hist.*, 42(2), 327–53.

Anderson, James N. (1964), 'Land and Society in a Pangasinan Society', in S. C. Espiritu and C. H. Hunt (eds.), *Foundation of Community Development*. Manila, Philippines: Garcia.

Aoki, Masahiko (ed.) (1984), *The Economic Analysis of the Japanese Firm*. Amsterdam: North-Holland.

Appu, P. S. (1975), 'Tenancy Reform in India', *Econ. Polit. Weekly*, 10(33–5), 1339–75.

Arrow, Kenneth J. (1968), 'The Economics of Moral Hazard: Further Comment', *Amer. Econ. Rev.*, 58(3), 537–9.

—— (1974), *Limits of Organization*. New York: Norton

—— (1985), 'The Economics of Agency', in J. Pratt and R. Zeckhauser (eds.), *Principals and Agents: The Structure of Business*. Boston: Harvard Business School Press, 37–51.

—— Chenery, H. B., Minhas, B. S., and Solow, R. M. (1961), 'Capital-Labor Substitution and Economic Efficiency', *Rev. Econ. Stat.*, 43(3), 225–50.

—— and Kurtz, Mordecai (1970), *Public Investment, the Rate of Return, and Optimal Fiscal Policy*. Baltimore: Johns Hopkins University Press.

Bagi, F. S. (1981), 'Economic Efficiency of Share-Cropping in Indian Agriculture', *Malayan Econ. Rev.*, 26(1), 15–24.

Bardhan, Kalpana (1977), 'Rural Employment, Wages, and Labor Markets in India: A Survey of Research-III', *Econ. Polit. Weekly*, 12(28), 1101–18.

Bardhan, Pranab K. (1977), 'Variations in Forms of Tenancy in a Peasant Economy', *J. Devel. Econ.*, 4(2), 105–18.

—— (1979a), 'Agricultural Development and Land Tenancy in a Peasant Economy: A Theoretical and Empirical Analysis', *Amer. J. Agr. Econ.*, 61(1), 48–57.

—— (1979b), 'Wages and Unemployment in a Poor Agrarian Economy: A Theoretical and Empirical Analysis', *J. Polit. Econ.*, 87(3), 479–500.

—— (1980), 'Interlocking Factor Markets and Agrarian Development', *Oxf. Econ. Pap.*, 32(1), 82–98.

—— (1983), 'Labor-Tying in a Poor Agrarian Economy: A Theoretical and Empirical Analysis', *Quart. J. Econ.*, 98(3), 501–14.

—— (1985), 'Agricultural Development and Land Tenancy in a Peasant Economy: Reply', *Amer. J. Agr. Econ.*, 67(3), 691–2.

—— (1984), *Land, Labor and Rural Poverty: Essays in Development Economics*. New York: Columbia University Press.

—— (ed.) (1989), *The Economic Theory of Agrarian Institutions*. Oxford: Clarendon Press.

—— and Rudra, Ashok (1980), 'Terms and Conditions of Sharecropping Contracts: An Analysis of Village Survey Data in India', *J. Devel. Stud.*, 16(3), 287–302.

—— —— (1981), 'Terms and Conditions of Labor Contracts in Agriculture: Results of a Survey in West Bengal, 1979', *Oxf. Bul. Econ. Stat.*, 43(1), 89–111.

—— and Singh, Nirvikar (1987), 'On Moral Hazard and Cost Sharing under Sharecropping', *Amer. J. Agr. Econ.*, 69(2), 382–3.

—— and Srinivasan, T. N. (1971), 'Cropsharing Tenancy in Agriculture: A Theoretical and Empirical Analysis', *Amer. Econ. Rev.*, 61(1), 48–64.

Barker, Randolph, and Cordova, V. G. (1978), 'Labour Utilization in Rice Production', in *Economic Consequences of the New Rice Technology*. Los Baños, Philippines: International Rice Research Institute.

—— and Herdt, Robert W. (1985), *The Rice Economy of Asia*. Washington, DC: Resources for the Future.

References

Basant, Rakesh (1984), 'Attached and Casual Labor Wage Rates', *Econ. Polit. Weekly*, 19(9), 390–6.

Basu, Kaushik (1983), 'The Emergence of Isolation and Interlinkage in Rural Markets', *Oxf. Econ. Pap.*, 35(2), 262–80.

—— (1984), 'Implicit Interest Rates, Usury and Isolation in Backward Agriculture', *Cambridge J. Econ.*, 8(2), 145–59.

—— (1987), 'Disneyland Monopoly, Interlinkage, and Usurious Interest Rates', *J. Pub. Econ.*, 34(1), 1–17.

Becker, Gary S. (1974), 'A Theory of Social Interactions', *J. Polit. Econ.*, 82(6), 1063–93.

—— (1976), 'Altruism, Egoism, and Genetic Fitness: Economics and Sociobiology', *J. Econ. Lit.*, 14(3), 817–26.

Bell, Clive (1977), 'Alternative Theories of Sharecropping: Some Tests Using Evidence from Northeast India', *J. Devel. Stud.*, 13(4), 317–46.

—— (1988), 'Credit Markets and Interlinked Transactions', in H. Chenery and T. N. Srinivasan (eds.), *Handbook of Development Economics*. Amsterdam: North-Holland, 763–830.

—— and Braverman, Avishay, (1980), 'On the Existence of "Marshallian" Sharecropping Contracts', *Indian Econ. Rev.*, 15(3), 201–3.

—— and Srinivasan, T. N. (1985a), 'Some Salient Features of Tenancy Contracts, Commodity Prices and Wages: A Comparison of Andhra Pradesh, Bihar, and Punjab.' Washington, DC: World Bank, mimeo.

—— (1985b), 'The Demand for Attached Farm Servants in Andhra Pradesh, Bihar, and Punjab.' Washington, DC: World Bank, mimeo.

—— (1989), 'Interlinked Transactions in Rural Markets: An Empirical Study of Andhra Pradesh, Bihar, and Punjab', *Oxf. Bul. Econ. Stat.*, 51(1), 73–83.

—— and Sussangkarn, Chalongphob (1985), 'The Choice of Tenancy Contracts'. Washington, DC: World Bank, mimeo.

—— and Zusman, Minhas (1976), 'A Bargaining Approach to Cropsharing Contracts', *Amer. Econ. Rev.*, 66(4), 578–88.

—— —— (1989), 'The Equilibrium Vector of Pairwise-Bargained Agency Contracts with Diverse Actors and Principals Owning a Fixed Resource', *J. Econ. Behav. Org.*, 11(1), 91–114.

Benoit, Jean-Pierre, and Krishna, Vijay (1985), 'Finitely Repeated Games', *Econometrica*, 53(4), 905–22.

Ben-Porath, Yoram (1980), 'The F-Connection: Families, Friends, and Firms and the Organization of Exchange', *Pop. Devel. Rev.*, 6(1), 1–30.

Berry, R. Albert, and Cline, William R. (1979), *Agrarian Structure and Productivity in Developing Countries*. Baltimore: Johns Hopkins University Press.

Berry, Russell L. (1962), 'Cost Sharing as a Means of Improving the Share Rent Lease', *J. Farm Econ.*, 44(3), 796–807.

Bester, Helmut (1985), 'Screening vs. Rationing in Credit Markets with Incomplete Information', *Amer. Econ. Rev.*, 75(4), 850–5.

Bhaduri, Amit (1973), 'A Study on Agricultural Backwardness under Semi-Feudalism', *Econ. J.*, 83(329), 120–37.

Bhagwati, Jagdish N. (1966), *The Economics of Underdeveloped Countries*. New York: McGraw-Hill.

—— and Chakravarty, S. (1969), 'Contributions to Indian Economic Analysis: A Survey', *Amer. Econ. Rev.*, 59 (Supplement), 1–73.

Bhalla, Sheila (1976), 'New Relations of Production in Haryana Agriculture', *Econ. Polit. Weekly*, 11(13), 23–30.

Bhalla, Surjit S., and Roy, Prannoy (1988), 'Mis-Specification in Farm Productivity Analysis: The Role of Land Quality', *Oxf. Econ. Pap.*, 40(1), 55–73.

Bharadwaj, Krishna (1974), *Production Conditions in Indian Agriculture: A Study based on Farm Management Surveys*. Cambridge: Cambridge University Press.

—— and Das, P. K. (1975), 'Tenurial Conditions and Mode of Exploitation: A Study of Some Villages in Orissa', *Econ. Polit. Weekly*, 10(5–7), 221–40.

Bhuiyan, M. S. R. (1987), 'Effect of Farm Size and Tenurial Status of Land on Production Efficiency in an Area of Bangladesh', *Bangladesh J. Agr. Econ.*, 10(1), 1–31.

—— and Nandal, D. S. (1987), 'Tenurial Status of Farm, Resource Endowment, Resource Use and Productive Efficiency in Mymensingh District of Bangladesh', *Indian J. Agr. Econ.*, 42(2), 207–20.

Binswanger, Hans P., Doherty, Victor S., Balaramaiah, T., Bhende, M. J., Kshirsagar, K. G., Rao, V. B., and Raju, P. S. S. (1984), 'Common Features and Contrasts in Labor Relations in the Semiarid Tropics of India', in H. P. Binswanger and M. R. Rosenzweig (eds.) *Contractual Arrangements, Employment, and Wages in Rural Labor Markets in Asia*. New Haven, Conn.: Yale University Press, 143–68.

—— and McIntire, John (1987), 'Behavioral and Material Determinants of Production Relations in Land Abundant Tropical Agriculture', *Econ. Develop. Cult. Change*, 36(1), 73–99.

—— and Rosenzweig, Mark R. (1984), 'Contractual Arrangements, Employment, and Wages in Rural Labor Markets: A Critical Review', in H. P. Binswanger and M. R. Rosenzweig (eds.), *Contractual Arrangements, Employment, and Wages in Rural Labor Markets in Asia*. New Haven, Conn.: Yale University Press, 1–40.

—— —— (1986), 'Behavioral and Material Determinants of Production Relations in Agriculture', *J. Devel. Stud.*, 22(3), 503–39.

—— and Sillers, D. A. (1983), 'Risk Aversion and Credit Constraints in Farmers' Decision-Making', *J. Devel. Stud.*, 20(1), 5–21.

Bliss, C. J., and Stern, N. H. (1982), *Palanpur: The Economy of an Indian Village*. New York: Clarendon Press.

Boyce, James K. (1987), *Agrarian Impasse in Bengal: Institutional Constraints to Technological Change*. Oxford: Oxford University Press.

Bradly, Michael E., and Clark, M. Gardner (1972), 'Supervision and Efficiency in Socialized Agriculture', *Soviet Stud.*, 23(4), 465–73.

Braverman, Avishay, and Guasch, J. Luis (1984), 'Capital Requirements, Screening, and Interlinked Sharecropping and Credit Contracts', *J. Devel. Econ.*, 14(3), 359–74.

—— —— (1986), 'Rural Credit Markets and Institutions in Developing Countries: Lessons for Policy Analysis from Practice and Modern Theory', *World Devel.*, 14(10/11), 1253–67.

—— and Srinivasan, T. N. (1981), 'Credit and Sharecropping in Agrarian Societies', *J. Devel. Econ.*, 9(3), 289–312.

—— and Stiglitz, Joseph E. (1982), 'Sharecropping and the Interlinking of Agrarian Markets', *Amer. Econ. Rev.*, 72(4), 695–715.

—— —— (1986a), 'Cost-Sharing Arrangements under Sharecropping: Moral Hazard, Incentive Flexibility, and Risk', *Amer. J. Agr. Econ.*, 68(3), 642–52.

—— —— (1986b), 'Landlords, Tenants and Technological Innovation', *J. Devel. Econ.*, 23(2), 313–32.

Breman, Jan (1974), *Patronage and Exploitation: Changing Agrarian Relations in South Gujurat, India*. Berkeley, Calif.: University of California Press.

Brewster, John M. (1950), 'The Machine Process in Agriculture and Industry', *J. Farm Econ.*, 32(1), 69–81.

Bromley, Daniel W. (1986), 'Natural Resources and Agricultural Development in the Tropics: Is Conflict Inevitable?' in Allen Maunder and Ulf Renborg (eds.), *Agriculture in Turbulent World Economy*. Aldershot: Gower House, 319–27.

Bull, Clive (1983), 'Implicit Contracts in the Absence of Enforcement and Risk Aversion', *Amer. Econ. Rev.*, 73(4), 658–71.

—— (1987), 'The Existence of Self-Enforcing Implicit Contracts', *Quart. J. Econ.*, 102(1), 147–59.

Cain, Mead (1981), 'Risk and Insurance: Perspectives on Fertility and Agrarian Change in India and Bangladesh', *Pop. Devel. Rev.*, 7(3), 435–74.

Caldwell, John C., Reddy, P. H., and Caldwell, Pat (1986), 'Periodic High Risk as a Cause of Fertility Decline in a Changing Rural Environment: Survival Strategies in the 1980–83 South Indian Drought', *Econ. Develop. Cult. Change*, 34(4), 677–701.

Calvo, Guillermo A., and Wellisz, Stanislaw (1978), 'Supervision, Loss of Control, and the Optimum Size of the Firm', *J. Polit. Econ.*, 86(5), 943–52.

—— —— (1979), 'Hierarchy, Ability, and Income Distribution', *J. Polit. Econ.*, 87(5), 991–1010.

Carmichael, H. Lorne (1983), 'The Agent–Agent Problem: Payment by Relative Output', *J. Labor Econ.*, 1(1), 50–65.

—— (1984), 'Reputation in the Labor Market', *Amer. Econ. Rev.*, 74(4), 713–25.

Carter, Michael R. (1987), 'Risk Sharing and Incentives in the Decollectivization of Agriculture', *Oxf. Econ. Pap.*, 39(3), 577–95.

—— (1988), 'Equilibrium Credit Rationing of Small Farm Agriculture', *J. Devel. Econ.*, 28(1), 83–103.

Castillo, Gelia T. (1975), *All in a Grain of Rice: A Review of Philippine Studies on the Social and Economic Implications of the New Rice Technology*. Laguna, Philippines: Southeast Asian Regional Center for Graduate Study and Research in Agriculture.

Castle, Emery (1952), 'Some Aspects of the Crop-Share Lease', *Land Econ.*, 28(2), 177–9.

Chandra, N. K. (1974), 'Farm Efficiency under Semi-Feudalism: A Critique of Marginalist Theories and Some Marxist Formulations', *Econ. Polit. Weekly*, 9(32–4), Special Number, 1309–32.

Chakravarty, Aparajita, and Rudra, Ashok (1973), 'Economic Effects of Tenancy: Some Negative Results', *Econ. Polit. Weekly*, 8(28), 1139–46.

Chao, Kang (1983), 'Tenure Systems in Traditional China', *Econ. Develop. Cult. Change*, 31(2), 295–314.

Chattopadhyay, Manabendu (1979), 'Relative Efficiency of Owner and Tenant Cultivation: A Case Study', *Econ. Polit. Weekly*, 14(39), A93–6.

Chayanov, Alexander V. (1966), *Theory of Peasant Economy*. Homewood, Ill.: Richard D. Irwin.

Cheung, Steven N. S. (1969), *The Theory of Share Tenancy*. Chicago: University of Chicago Press.

Chuma, Hiroyuki, Otsuka, Keijiro, and Hayami, Yujiro (1990), 'On the Dominance of Tenancy over Permanent Labor Contract in Agrarian Economies', *J. Japanese and International Economies*, 4(2), 101–20.

Clayton, Eric S. (1964), *Agrarian Development in Peasant Economies: Some Lessons from Kenya*. Oxford: Pergamon.

Coase, Ronald H. (1937), 'The Nature of the Firm', *Economica*, 4(16), 386–405.

—— (1960), 'The Problem of Social Cost', *J. Law. Econ.*, 3(1), 1–44.

Cohen, P. T. (1983), 'Problems of Tenancy and Landlessness in Northern Thailand', *Developing Economies*, 21(3), 244–66.

Currie, J. M. (1981), *The Economic Theory of Agricultural Land Tenure*. Cambridge: Cambridge University Press.

Dantwala, M. L., and Shah, C. H. (1971), *Evaluation of Land Reforms*, i, Bombay: University of Bombay Press.

Datta, Samar K., and Nugent, Jeffrey B. (1985), 'Agricultural Development and Land Tenancy in a Peasant Economy: Comment', *Amer. J. Agr. Econ.*, 67(3), 688–90.

—— —— Tishler, Asher, and Wang, Jonelin (1988), 'Seasonality, Differential Access and Interlinking of Labor and Credit', *J. Devel. Stud.*, 24(3), 379–93.

—— O'Hara, Donald J., and Nugent, Jeffrey B. (1986), 'Choice of Agricultural Tenancy in the Presence of Transaction Costs', *Land Econ.*, 62(2), 145–58.

David, Cristina C., and Otsuka, Keijiro (1990), 'The Modern Seed-Fertilizer Technology and Adoption of Labor-Saving Technologies: The Philippine Case', *Australian J. Agr., Econ.*, 34(2), 132–46.

Day, Richard J. (1967), 'The Economics of Technological Change and the Demise of the Sharecropper', *Amer. Econ. Rev.*, 57(3), 427–49.

de Janvry, Alain (1981), *The Agrarian Question and Reformism in Latin America*. Baltimore: Johns Hopkins University Press.

—— Fukui, Seiichi, and Sadoulet, Elisabeth (1989), 'Efficient Share Tenancy Contracts under Risk: The Case of Three Rice-Growing Villages in Thailand'. Berkeley, Calif.: University of California, mimeo.

de los Reyes, B. N. (1972), 'Can Land Reform Succeed?' *Philippine Sociological Rev.*, 20(1), 79–92.

Dow, N. (1984), 'Usufruct and Usury: An Analysis of Land Leasing in East Java', *Australian J. Agr. Econ.*, 28(2), 15–32.

Dowell, Richard S. (1977), 'Risk Diversification and Land Tenure in the United States Agriculture', Ph.D. Diss.. University of Chicago, 1977.

Drake, Louis S. (1952), 'Comparative Productivity of Share- and Cash-Rent Systems of Tenure', *J. Farm Econ.*, 34(4), 535–50.

Dwivedi, Harendranath, and Rudra, Ashok (1973), 'Economic Effects of Tenancy: Some Further Negative Results', *Econ. Polit. Weekly*, 8(29), 1291–4.

Ely, Richard T., and Galpin, Charles J. (1919), 'Tenancy in an Ideal System of Land Ownership', *Amer. Econ. Rev.*, 9(1), Supplement, 180–212.

Embree, John F. (1950), 'Thailand: A Loosely Structured Social System', *Amer. Anthropologist*, 52(1), 181–93.

Eswaran, Mukesh, and Kotwal, Ashok (1985a), 'A Theory of Two-Tier Labor Markets in Agrarian Economies', *Amer. Econ. Rev.*, 75(1), 162–77.

—— —— (1985b), 'A Theory of Contractual Structure in Agriculture', *Amer. Econ. Rev.*, 75(3), 352–77.

—— —— (1986), 'Access to Capital and Agrarian Production Organization', *Econ. J.*, 96(382), 482–98.

—— —— (1989), 'Credit as Insurance in Agrarian Economies', *J. Devel. Econ.*, 31(1), 37–53.

Feder, Gershon (1985), 'The Relation between Farm Size and Farm

Productivity: The Role of Family Labor, Supervision, and Credit Constraints', *J. Devel. Econ.*, 18(2–3), 297–313.

—— Just, Richard E., and Zilberman, David (1985), 'Adoption of Agricultural Innovation in Developing Countries: A Survey', *Econ. Develop. Cult. Change*, 33(2), 255–98.

—— Onchan, Tongroj, Chalamwong, Yongyuth, and Hongladarom, Chira (1988), *Land Policies and Farm Productivity in Thailand*. Baltimore: Johns Hopkins University Press.

Finkler, Kaja (1978), 'From Sharecroppers to Entrepreneurs: Peasant Household Production Strategies under Ejido Systems of Mexico', *Econ. Develop. Cult. Change*, 27(1), 103–20.

Friedman, Milton (1953), 'The Methodology of Positive Economics', in *Essays in Positive Economics*. Chicago: University of Chicago Press.

Fujimoto, Akimi (1983), *Income Sharing among Malay Peasants: A Study of Land Tenure and Rice Production*. Singapore: Singapore University Press.

—— (1985), 'Land Tenure and Tenancy Relations in a Subang Village in West Java', in A. Fujimoto and T. Matsuda (eds.), *A Comparative Study of the Structure of Rice Productivity and Rural Society: Two Village Studies in Indonesia and Thailand*. Tokyo, Japan: Tokyo University of Agriculture, 51–67.

—— (1986), 'Share Tenancy and Rice Production: Lessons from Two Village Studies in West Java', in A. Fujimoto and T. Matsuda (eds.), *An Economic Study of Rice Farming in West Java*. Tokyo, Japan: Tokyo University of Agriculture, 81–99.

Furnival, John A. (1944), *Netherlands India: A Study of Plural Economy*. Cambridge: Cambridge University Press.

Gangopadhyay, Shubhashis, and Sengupta, Kunal (1986), 'Interlinkages in Rural Markets', *Oxf. Econ. Pap.*, 38(1), 112–21.

—— (1987), 'Small Farmers, Moneylenders, and Trading Activity', *Oxf. Econ. Pap.*, 39(2), 333–42.

Gapud, Jose P. (1959), 'Financing Lowland Rice Farming in Selected Barrios of Muñoz, Nueva Ecija', *Econ. Res. J.*, 6(1), 74–82.

Geertz, Clifford (1970), *Agricultural Involution: The Process of Ecological Change in Indonesia*. Berkeley and Los Angeles: University of California Press.

—— (1978), 'The Bazaar Economy: Information and Search in Peasant Marketing', *Amer. Econ. Rev.*, 68(2), 28–32.

George, Alex (1987), 'Social and Economic Aspects of Attached Laborers in Kuttanad Agriculture', *Econ. Polit. Weekly*, 22(52), A141–50.

Georgescu-Roegen, Nicholas (1960), 'Economic Theory and Agrarian Economics', *Oxf. Econ. Pap.*, 12(1), 1–40.

Ghose, A. K. (1980), 'Wages and Employment in Indian Agriculture', *World Devel.*, 8(5 and 6), 413–28.

References

Green, Jerry R., and Stokey, Nancy L. (1983), 'A Comparison of Tournaments and Contracts', *J. Polit. Econ.*, 91(3), 349–64.

Guhan, S., and Bharathan, K. (1984), 'Dusi: A Resurvey', Working Paper No. 52. Madras: Madras Institute of Development Studies.

Hallagan, William (1978), 'Self-Selection by Contractual Choice and the Theory of Sharecropping', *Bell J. Econ.*, 9(2), 344–54.

Harris, Milton, and Raviv, Artur (1978), 'Some Results on Incentive Contracts with Applications to Education and Employment, Health Insurance, and Law Enforcement', *Amer. Econ. Rev.*, 68(1), 20–30.

—— —— (1979), 'Optimal Incentive Contracts with Imperfect Information', *J. Econ. Theory*, 20(2), 231–59.

Hart, Gillian (1986), 'Interlocking Transactions: Obstacles, Precursors or Instruments of Agrarian Capitalism?' *J. Devel. Econ.*, 23(1), 173–203.

Hart, Oliver, and Holmstrom, Bengt (1987), 'The Theory of Contracts', in T. Bewley (ed.), *Advances in Economic Theory*. Cambridge: Cambridge University Press, 71–155.

Hayami, Yujiro (1990), 'Community, Market, and State', Elmhirst Memorial Lecture, in A. Maunder and A. Valdes (eds.), *Agriculture and Governments in an Interdependent World*. Aldershot: Dartmouth, 3–14.

—— and Kawagoe, Toshihiko (1989), 'Farm Mechanization, Scale Economies and Polarization: The Japanese Experience', *J. Devel. Econ.*, 31(2), 221–39.

—— and Kikuchi, Masao (1982), *Asian Village Economy at the Crossroads*. Baltimore: Johns Hopkins University Press.

—— —— Moya, P. F., Bambo, L. M., and Marciano E. B. (1978), *Anatomy of a Peasant Economy: A Rice Village in the Philippines*. Los Baños, Philippines: International Rice Research Institute.

—— and Otsuka, Keijiro (1992), '*Kasugpong* in the Philippine Rice Bowl: The Emergence of New Labor Institutions after the Land Reform', in A. Braverman, K. Hoff, and J. E. Stiglitz (eds.), *Agricultural Development Policies and the Theory of Rural Organization*. Oxford: Oxford University Press, forthcoming.

—— Quisumbing, M. Agnes, and Adriano, Lourdes S. (1990), *Toward an Alternative Land Reform Paradigm: A Philippine Perspective*. Quezon City, Philippines: Ateneo de Manila University Press.

—— and Ruttan, Vernon W. (1985), *Agricultural Development: An International Perspective*. Rev. edn. Baltimore: Johns Hopkins University Press.

Heady, Earl O. (1947) 'Economics of Farm Leasing Systems', *J. Farm Econ.*, 29(3), 659–78.

—— (1955), 'Marginal Resource Productivity and Imputation of Shares for a Sample of Rented Farms', *J. Polit. Econ.*, 63(6), 500–11.

Hendry, James B. (1960), 'Land Tenure in South Vietnam', *Econ. Develop. Cult. Change*, 9(1), 27–44.

Herdt, Robert W. (1978), 'Costs and Returns for Rice Production', in *Economic Consequences of the New Rice Technology*. Los Baños, Philippines: International Rice Research Institute.

—— (1987), 'A Retrospective View of Technological and Other Changes in Philippine Rice Farming, 1965–1982', *Econ. Develop. Cult. Change*, 35(2), 329–49.

Herring, Ronald J. (1983), *Land to the Tiller: The Political Economy of Agrarian Reforms in South Asia*. New Haven, Conn.: Yale University Press.

Hester, Evett D., and Mabun, Pablo (1924), 'Some Economic and Social Aspects of Philippine Tenancies', *Philippine Agriculturist*, 12, 367–444.

Hiebert, L. Dean (1978), 'Uncertainty and Incentive Effects in Share Contracts', *Amer. J. Agr. Econ.*, 60(3), 536–9.

Higgs, Henry (1894), ' "Métayage" in Western France', *Econ. J.*, 4(1), 1–13.

Higgs, Robert (1973), 'Race, Tenure, and Resource Allocation in Southern Agriculture, 1910', *J. Econ. Hist.*, 33(1), 149–69.

—— (1974), 'Patterns of Farm Rental in the Georgia Cotton Belt, 1880–1900', *J. Econ. Hist.*, 34(2), 468–82.

Hirashima, Shigemochi (ed.) (1977), *Hired Labor in Rural Asia*. Tokyo, Japan: Institute of Developing Economies.

—— (1978), *The Structure of Disparity in Development: The Case Study of the Pakistan Punjab Agriculture*. Tokyo, Japan: Institute of Developing Economies.

Hirschmeir, Johannes, and Yui, Tsunehiko (1981), *The Development of Japanese Business*. London: Allen & Unwin.

Hirshleifer, Jack, and Riley, John G. (1979), 'The Analytics of Uncertainty and Information: An Expository Survey', *J. Econ. Lit.*, 17(4), 1375–421.

Ho, Samuel P. S. (1976), 'Uncertainty and the Choice of Tenure Arrangements: Some Hypotheses', *Amer. J. Agr. Econ.*, 58(1), 88–92.

Hoffman, Philip T. (1982), 'Sharecropping and Investment in Agriculture in Early Modern France', *J. Econ. Hist.*, 42(1), 155–62.

—— (1984), 'The Economic Theory of Sharecropping in Early Modern France', *J. Econ. Hist.*, 44(2), 309–19.

Holmstrom, Bengt (1979), 'Moral Hazard and Observability', *Bell J. Econ.*, 10(1), 74–91.

—— (1981), 'Contractual Models of the Labor Market', *Amer. Econ. Rev.*, 71(2), 308–13.

—— (1982), 'Moral Hazard in Teams', *Bell J. Econ.*, 13(2), 324–40.

—— (1983), 'Equilibrium Long Term Labor Contracts', *Quart. J. Econ.*, Supplement, 98(5), 23–54.

—— and Myerson, Roger B. (1983), 'Efficient and Durable Decision Rules with Incomplete Information', *Econometrica*, 51(6), 1789–819.

Horii, Kenzo (1981), *Rice Economy and Land Tenure in West Malaysia*. Tokyo, Japan: Institute of Developing Economies.

Hossain, Mahabub (1977), 'Farm Size, Tenancy and Land Productivity: An Analysis of Farm Level Data in Bangladesh Agriculture', *Bangladesh Devel. Stud.*, 5(3), 285–348.

—— (1978), 'Factors Affecting Tenancy: The Case of Bangladesh Agriculture', *Bangladesh Devel. Stud.*, 6(2), 139–62.

—— and Akash, M. M. (1990), 'Modern Rice Technology and Income Adjustment through Factor Markets: The Bangladesh Case', paper presented at the Final Workshop on the Differential Impact of the Modern Rice Technology across Production Environments. Los Baños, Philippines: International Rice Research Institute.

Hsiao, James C. (1975), 'The Theory of Share Tenancy Revisited', *J. Polit. Econ.*, 83(5), 1023–32.

Huang, Yukon (1975), 'Tenancy Patterns, Productivity, and Rentals in Malaysia', *Econ. Develop. Cult. Change*, 23(4), 703–18.

Hurwicz, Leonid, and Shapiro, Leonard (1978), 'Incentive Structures Maximizing Residual Gain under Incomplete Information', *Bell J. Econ.*, 9(1), 180–91.

Husken, Frans (1979), 'Landlords, Sharecroppers and Agricultural Laborers: Changing Labor Relations in Rural Java', *J. Contemporary Asia*, 9(2), 140–51.

Issawi, Charles (1957), 'Farm Output under Fixed Rents and Share Tenancy', *Land Econ.*, 33(1), 74–7.

Jabbar, M. A. (1977), 'Relative Productive Efficiency of Different Tenure Classes in Selected Areas of Bangladesh', *Bangladesh Devel. Stud.*, 5(1), 17–50.

James, William, and Roumasset, James (1984), 'Migration and the Evolution of Tenure Contracts in Newly Settled Regions', *J. Devel. Econ.*, 14(1), 147–62.

Jannuzi, F. Tomasson, and Peach, James T. (1980), *The Agrarian Structure of Bangladesh*. Boulder, Col.: Westview.

Jaynes, Gerald David (1982), 'Production and Distribution in Agrarian Economies', *Oxf. Econ. Pap.*, 34(2), 346–67.

—— (1984), 'Economic Theory and Land Tenure', in H. P. Binswanger and M. R. Rosenzweig (eds.), *Contractual Arrangements, Employment, and Wages in Rural Labor Markets in Asia*. New Haven, Conn.: Yale University Press, 43–62.

Jewitt, Ian (1988), 'Justifying the First-Order Approach to Principal-Agent Problems', *Econometrica*, 56(5), 1177–90.

Jodha, N. S. (1984), 'Agricultural Tenancy in Semiarid Tropical India', in H. P. Binswanger and M. R. Rosenzweig (eds.), *Contractual Arrangements, Employment, and Wages in Rural Labor Markets in Asia*. New Haven, Conn.: Yale University Press, 96–113.

Johnson, D. Gale (1950), 'Resource Allocation under Share Contracts', *J. Polit. Econ.*, 58(2), 111–23.

Junankar, P. N. (1976), 'Land Tenure and Indian Agricultural Productivity', *J. Devel. Stud.*, 13(1), 42–60.

Katwal, B. B. (1986), 'Wages and Welfare: The Case of Attached vs. Casual Labor in the Nepal Tarai', Research Paper Series No. 31, HMG-USAID-GTZ-Winrock Project. Kathmandu: Nepal.

Kawagoe, Toshihiko, Hayami, Yujiro, and Ruttan, Vernon W. (1985), 'Intercountry Agricultural Production Function and Productivity Differences among Countries', *J. Devel. Econ.*, 19(1–2), 113–32.

Khandker, Shahidur R., Mestelman, Stuart, and Feeny, David (1987), 'Allocative Efficiency, the Aggregation of Labor Inputs, and the Effects of Farm Size and Tenancy Status: Tests from Rural Bangladesh', *J. Devel. Stud.*, 24(1), 31–42.

Khasnabis, Ratan, and Chakravarty, Jyotiprakash (1982), 'Tenancy, Credit, and Agrarian Backwardness: Results of a Field Survey', *Econ. Polit. Weekly*, 17(3), A21–32.

Khusro, A. M. (1969), 'Farm Size and Land Tenure in India', *Indian Econ. Rev.*, 4(2), 123–45.

Kikuchi, Masao, and Hayami, Yujiro (1980), 'Technology and Labor Contract: Two Systems of Rice Harvesting in the Philippines', *J. Comp. Econ.*, 4(4), 357–77.

Kislev, Yoav, and Peterson, Willis (1982), 'Prices, Technology, and Farm Size', *J. Polit. Econ.*, 90(3), 578–95.

Kmenta, J. (1971), *Elements of Econometrics*. New York: Macmillan.

Knowles, James C., and Anker, Richard (1981), 'An Analysis of Income Transfers in a Developing Country', *J. Devel. Econ.*, 8(2), 205–26.

Koo, Anthony Y. C. (1973), 'Toward a More General Model of Land Tenancy and Reform', *Quart. J. Econ.*, 87(4), 567–80.

—— (1977), 'An Economic Justification for Land Reformism', *Econ. Develop. Cult. Change*, 25(3), 523–38.

—— (1982), *Land Market Distortion and Tenure Reform*. Ames, Ia.: Iowa State University Press.

Kotwal, Ashok (1985), 'The Role of Consumption Credit in Agricultural Tenancy', *J. Devel. Econ.*, 18(2 and 3), 273–95.

Kuchiba, Masuo, and Bauzon, Leslie E. (1979), *A Comparative Study of Paddy-Growing Communities in South and Southeast Asia and Japan*. Tokyo, Japan: Toyota Foundation.

—— Tsubouchi, Yoshihiro, and Maeda, Naribumi (eds.) (1979), *Three Malay Villages: A Sociology of Paddy Growers in West Malaysia*. Honolulu: University Press of Hawaii.

Kutcher, Gary P., and Scandizzo, Pasquale L. (1976), 'A Partial Analysis of Share-Tenancy Relationships in Northeast Brazil', *J. Devel. Econ.*, 3(4), 343–54.

References

Ladejinsky, Wolf (1977), *Agrarian Reform as Unfinished Business: Selected Papers of Wolf Ladejinsky*, edited by Louis J. Walinsky. Oxford: Oxford University Press.

Lambert, Richard A. (1983), 'Long-Term Contracts and Moral Hazard', *Bell J. Econ.*, 14(2), 441–52.

Lau, Lawrence J., and Yotopoulos, Pan A. (1989), 'The Meta-Production Function Approach to Technological Change in World Agriculture', *J. Devel. Econ.*, 31(2), 241–69.

Lazear, Edward P., and Rosen, Sherwin (1981), 'Rank-Order Tournaments as Optimum Labor Contracts', *J. Polit. Econ.*, 89(5), 841–64.

Ledesma, Antonio J. (1982), *Landless Workers and Rice Farmers: Peasant Subclasses under Agrarian Reform in Two Philippine Villages*. Los Baños, Philippines: International Rice Research Institute.

Lehmann, David (1986), 'Sharecropping and the Capitalist Transition in Agriculture: Some Evidence from the Highlands of Ecuador', *J. Devel. Econ.*, 23(2), 333–54.

Leibenstein, Harvey (1987), *Inside the Firm: The Inefficiency of Hierarchy*. Cambridge, Mass.: Harvard University Press.

Levinthal, Daniel (1988), 'A Survey of Agency Models of Organizations', *J. Econ. Behav. Org.*, 9(2), 153–85.

Lewis, Tracy R. (1980), 'Bonuses and Penalties in Incentive Contracting', *Bell J. Econ.*, 11(1), 292–301.

Lewis, W. Arthur (1969), *Aspects of Tropical Trade, 1883–1965*. Stockholm, Sweden: Almquist & Wiksel.

Lin, Justin Y. (1988), 'The Household Responsibility System in China's Agricultural Reform: A Theoretical and Empirical Study', *Econ. Develop. Cult. Change*, 36(3), Supplement, 199–224.

Lowe, R. (1986), *Agricultural Revolution in Africa*. London: Macmillan.

Lucas, Robert E. B. (1979), 'Sharing, Monitoring, and Incentives: Marshallian Misallocation Reassessed', *J. Polit. Econ.*, 87(3), 501–21.

—— and Stark, Oded (1985), 'Motivations to Remit: Evidence from Botswana', *J. Polit. Econ.*, 93(5), 901–18.

Luce, R. Duncan, and Raiffa, Howard (1957), *Games and Decisions*. New York: John Wiley & Sons.

Mandal, M. A. S. (1980), 'Farm Size, Tenure, and Productivity in an Area of Bangladesh', *Bangladesh J. Agr. Econ.*, 3(2), 21–42.

Mangahas, Mahar (1975), 'An Economic Theory of Tenant and Landlord Based on a Philippine Case', in L. G. Reynolds (ed.), *Agriculture in Development Theory*. New Haven, Conn.: Yale University Press, 138–61.

—— (1985), 'Rural Poverty and Operation Land Transfer in the Philippines', in Riswanul Islam (ed.), *Strategies for Alleviating Poverty in Rural Asia*. Dhaka: Bangladesh Institute of Development Studies.

—— Miralao, Virginia A., and de los Reyes, Romana P. (1976), *Tenants,*

Lessees, Owners: Welfare Implications of Tenure Change. Quezon City, Philippines: Ateneo de Manila University Press.

Marshall, Alfred (1956), *Principles of Economics*, 8th edn., London: Macmillan [1890].

Mazumdar, Dipak (1975), 'The Theory of Share-Cropping with Labor Market Dualism', *Economica*, 42(67), 261–71.

McLennan, Marshall S. (1969), 'Land and Tenancy in the Central Luzon Plain', *Philippine Stud.*, 17, 651–82.

—— (1982), 'Changing Human Ecology in the Central Luzon Plain: Nueva Ecija, 1709–1935', in A. W. McCoy and E. C. de Jesus (eds.), *Philippine Social History: Global Trend and Local Transformations*. Manila, Philippines: Ateneo de Manila University Press.

Milgrom, Paul, and Stokey, Nancy (1982), 'Information, Trade, and Common Knowledge', *J. Econ. Theory*, 26(1), 17–27.

Mill, John Stuart (1926), *Principles of Political Economy*. Ashley edn., London: Longmans [1848].

Mirrlees, James A. (1974), 'Notes on Welfare Economics, Information and Uncertainty', in M. S. Balch, D. L. McFadden, and S. Y. Wu (eds.), *Essays on Economic Behaviour under Uncertainty*. Amsterdam, Netherlands: North-Holland, 243–61.

Mitra, Pradeep K. (1983), 'A Theory of Interlinked Rural Transactions', *J. Pub. Econ.*, 20(2), 167–91.

Mitrany, David (1951), *Marx against the Peasant*. Durham, NC: University of North Carolina Press.

Moffet, Dennis (1978), 'A Note on the Yaari Life Cycle Model', *Rev. Econ. Stud.*, 45(2), 385–8.

Morooka, Yoshinori, and Hayami, Yujiro (1989), 'Contract Choice and Enforcement in an Agrarian Community: Agricultural Tenancy in Upland Java', *J. Devel. Stud.*, 26(1), 28–42.

—— Mayrowani, Henny, Yuyus, Rusmiati, Yayah, Rokayah, and Kosugi, Sho (1989), *Soybean-based Farming System in Upland Java*. Bogor, Indonesia: ESCAP-CGPRT Centre.

Mundlak, Yair (1961), 'Empirical Production Function Free of Management Bias', *J. Farm Econ.*, 43(1), 44–56.

Murrell, Peter (1983), 'The Economics of Sharing: A Transaction Cost Analysis of Contractual Choice in Farming', *Bell J. Econ.*, 14(1), 283–93.

Myerson, Roger B. (1979), 'Incentive Compatibility and Bargaining Problem', *Econometrica*, 47(1), 61–73.

Nabi, Ijaz (1986), 'Contracts, Resource Use and Productivity in Sharecropping', *J. Devel. Stud.*, 22(2), 429–42.

Narain, Dharm, and Joshi, P. C. (1969), 'Magnitude of Agricultural Tenancy', *Econ. Polit. Weekly*, 4(39), A139–42.

Nelson, Joan M. (1976), 'Sojourners versus New Urbanites: Causes and

Consequences of Temporary versus Permanent Cityward Migration in Developing Countries', *Econ. Develop. Cult. Change*, 24(4), 721–57.
Newbery, David M. G. (1974), 'Cropsharing Tenancy in Agriculture: A Comment', *Amer. Econ. Rev.*, 64(6), 1060–6.
—— (1975a), 'The Choice of Rental Contract in Peasant Agriculture', in Lloyd G. Reynolds (ed.), *Agriculture in Development Theory*. New Haven, Conn.: Yale University Press, 109–37.
—— (1975b), 'Tenurial Obstacles to Innovation', *J. Devel. Stud.*, 11(4), 263–77.
—— (1977), 'Risk Sharing, Sharecropping, and Uncertain Labor Markets', *Rev. Econ. Stud.*, 44(4), 585–94.
—— and Stiglitz, Joseph E. (1979), 'Sharecropping, Risk Sharing, and the Importance of Imperfect Information', in J. A. Roumasset, J. M. Boussard, and I. Singh (eds.), *Risk, Uncertainty, and Agricultural Development*. Laguna, Philippines: Agricultural Development Council and Southeast Asian Regional Center for Graduate Study and Research in Agriculture, 311–39.
Oishi, Shinzaburo (1958), *Hokenteki Tochishovy no Kaitai Katei* (The Process of Dissolution of Feudal Land Ownership). Tokyo: Ochanomizu Shobo.
Otsuka, Keijiro (1991), 'Determinants and Consequences of Land Reform Implementation in the Philippines', *J. Devel. Econ.*, 35(2), 339–55.
—— and Hayami, Yujiro (1988), 'Theories of Share Tenancy: A Critical Survey', *Econ. Develop. Cult. Change*. 37(1), 31–68.
—— and Murakami, Naoki (1987), 'Resource Allocation and Efficiency of Sharecropping under Uncertainty', *Asian Econ. J.*, 1(1), 125–45.
—— —— (1989), 'Incentives and Enforcement under Contract: The Taxicab in Kyoto', *J. Japanese and International Economies*, 3(3), 231–49.
—— Chuma, Hiroyuki, and Hayami, Yujiro (1992), 'Land and Labor Contracts in Agrarian Economies: Theories and Facts', *J. Econ. Lit.*, forthcoming.
—— —— —— (1993), 'Permanent Labor and Land Tenancy Contracts in Agrarian Economies: An Integrated Analysis', *Economica*, forthcoming.
—— Cordova, Violeta G., and David, Cristina C. (1992), 'Green Revolution, Land Reform, and Household Income Distribution in the Philippines', *Econ. Develop. Cult. Change*, forthcoming.
—— Kikuchi, Masao, and Hayami, Yujiro (1986), 'Community and Market in Contract Choice: Jeepney in the Philippines', *Econ. Develop. Cult. Change*, 34(2), 279–98.
—— Marciano, Esther, Palis, Dolor, and Hayami, Yujiro (1989), 'Modern Rice Technology, Land Reform, and Agrarian Contracts: The Case of Central Luzon', Department of Agricultural Economics Paper No. 89–23. Los Baños, Philippines: International Rice Research Institute.
Pal, T. K. (1975), 'Cuttack, Orissa'. in *Changes in Rice Farming in*

Selected Areas of Asia. Los Baños, Philippines: International Rice Research Institute, 133–48.

Pant, Chandrashekar (1983), 'Tenancy and Family Resources', *J. Devel. Econ.*, 12(10), 27–39.

Parthasarathy, G. (1975), 'West Godavari, Andhra Pradesh', in *Changes in Rice Farming in Selected Areas of Asia*. Los Baños, Philippines: International Rice Research Institute.

—— and Prasad, D. S. (1974), 'Responses to, and Impact of, HYV Rice according to Land Size and Tenure in a Delta Village, Andhra Pradesh, India', *Developing Economies*, 12(2), 182–98.

Platteau, Jean-Philippe, and Abraham, Anita (1987), 'An Inquiry into Quasi-Credit Contracts: The Role of Reciprocal Credit and Interlinked Deals in Small-Scale Fishing Communities', *J. Devel. Stud.*, 23(4), 461–90.

Pollak, Robert (1985), 'A Transaction Cost Approach to Families and Households', *J. Econ. Lit.*, 23(2), 581–608.

Posner, Richard A. (1980), 'A Theory of Primitive Society with Special Reference to Law', *J. Law. Econ.*, 23(1), 1–53.

Prosterman, Roy L., and Riedinger, Jeffrey M. (1987), *Land Reform and Democratic Development*. Baltimore: Johns Hopkins University Press.

Quibria, M. G., and Rashid, Salim (1984), 'The Puzzle of Sharecropping: A Survey of Theories', *World Devel.*, 12(2), 103–14.

—— —— (1986), 'Sharecropping in Dual Agrarian Economies: A Synthesis', *Oxf. Econ. Pap.*, 38(1), 94–111.

Radner, Roy (1981), 'Monitoring Cooperative Agreements in a Repeated Principal-Agent Relationship', *Econometrica*, 49(5), 1127–48.

Raj, K. N. (1970), 'Ownership and Distribution of Land', *Indian Econ. Rev.*, 5(1), 1–37.

Ransom, Roger L., and Sutch, Richard (1973), 'The Ex-Slave in the Post-Bellum South: A Study of the Economic Impact of Racism in a Market Environment', *J. Econ. Hist.*, 33(1), 131–48.

Rao, C. H. Hanumantha (1971), 'Uncertainty, Entrepreneurship, and Sharecropping in India', *J. Polit. Econ.*, 79(3), 578–95.

—— (1975), *Technological Change and Distribution of Gains in Indian Agriculture*. Delhi, India: Macmillan Co. of India.

Rao, J. Mohan (1986), 'Agriculture in Recent Development Theory', *J. Devel. Econ.*, 22(1), 41–86.

—— (1987), 'Productivity and Distribution under Cropsharing Tenancy', *World Devel.*, 15(9), 1163–78.

Ray, Debraj, and Sengupta, Kunal (1989), 'Interlinkages and the Pattern of Competition', in Bardhan (1989), 243–63.

Reid, Joseph, D., Jr. (1973), 'Sharecropping as an Understandable Market Response: The Post-Bellum South', *J. Econ. Hist.*, 33(1), 106–30.

—— (1975), 'Sharecropping in History and Theory', *Agr. Hist.*, 49(2), 426–40.
—— (1976a), 'Antebellum Southern Rental Contracts', *Exploration. Econ. Hist.*, 13(1), 69–83.
—— (1976b), 'Sharecropping and Agricultural Uncertainty', *Econ. Develop. Cult. Change*, 24(3), 549–76.
—— (1979a), 'White Land, Black Labor, and Agricultural Stagnation', *Exploration. Econ. Hist.*, 16(1), 31–55.
—— (1979b), 'Sharecropping and Tenancy in American History', in J. A. Roumasset, J. M. Boussard, and I. Singh (eds.), *Risk, Uncertainty and Agricultural Development*. Laguna, Philippines: Agricultural Development Council and Southeast Asian Regional Center for Graduate Study and Research in Agriculture, 283–309.
Richards, Alan (1979), 'The Political Economy of Gutswirtschaft: A Comparative Analysis of East Elbian Germany, Egypt, and Chile', *Compar. Stud. in Society and Hist.*, 21(4), 483–518.
Rivera, Generoso E., and McMillan, Robert T. (1954), *An Economic and Social Survey of Rural Households in Central Luzon*. Manila, Philippines: Cooperative Research Project of the Philippine Council for United States Aid and the United States of America Operations Mission to the Philippines.
Robertson, A. F. (1982), 'Abusa: The Structural History of an Economic Contract', *J. Devel. Stud.*, 18(4) 447–78.
—— (1987), *The Dynamics of Productive Relationships: African Share Contracts in Comparative Perspective*. Cambridge: Cambridge University Press.
Rosenzweig, Mark R. (1988a), 'Risk, Private Information, and the Family', *Amer. Econ. Rev.*, 78(2), 245–50.
—— (1988b), 'Risk, Implicit Contracts and the Family in Rural Areas of Low Income Countries', *Econ. J.*, 98(393), 1148–70.
—— (1988c), 'Labor Markets in Low Income Countries: Distortions, Mobility, and Migration', in H. Chenery and T. N. Srinivasan (eds.), *Handbook of Development Economics*. Amsterdam, Netherlands: North-Holland, 713–62.
—— and Stark, Oded (1989), 'Consumption Smoothing, Migration and Marriage: Evidence from Rural India', *J. Polit Econ.*, 97(4), 905–26.
—— and Wolpin, Kenneth J. (1985), 'Specific Experience, Household Structure and Intergenerational Transfers: Farm Family Land and Labor Arrangements in Developing Countries', *Quart. J. Econ.*, Suppl., 100, 961–88.
Roumasset, James A. (1976), *Rice and Risk: Decision Making among Low Income Farmers*. Amsterdam, Netherlands: North-Holland.
—— (1979), 'Sharecropping, Production Externalities and the Theory of Contracts', *Amer. J. Agr. Econ.*, 61(4), 640–7.

Roumasset, James A. (1984), 'Explaining Patterns in Landowner Shares: Rice, Corn, and Abaca in the Philippines', in H. P. Binswanger and M. R. Rosenzweig (eds.), *Contractual Arrangements, Employment, and Wages in Rural Labor Markets in Asia*. New Haven, Conn.: Yale University Press, 82–95.

—— and James, William T. (1979), 'Explaining Variations in Share Contracts: Land Quality, Population Pressure and Technological Change', *Australian J. Agr. Econ.*, 23(2), 116–27.

—— and Uy, Marilou (1980), 'Piece Rates, Time Rates, and Teams: Explaining Patterns in Employment Relation', *J. Econ. Behav. Org.*, 1(1), 343–60.

Rubinstein, Ariel, and Yaari, Menaham E. (1983), 'Repeated Insurance Contracts and Moral Hazard', *J. Econ. Theory*, 30(1), 74–97.

Rudra, Ashok (1971), 'Employment Patterns in Large Farms of Punjab', *Econ. Polit. Weekly*, 6(26), A89–94.

—— (1975), 'Loan as a Part of Agrarian Relations: Some Results of a Preliminary Survey in West Bengal', *Econ. Polit. Weekly*, 10(28), 1049–53.

Russel, Raymond (1985), 'Employee Ownership and Internal Governance', *J. Econ. Behav. Org.*, 6(3), 217–41.

Ruttan, Vernon W. (1964), 'Equity and Productivity Objectives in Agrarian Reform Legislation: Perspectives on the New Philippine Land Reform Code', *Indian J. Agr. Econ.*, 19(3), 115–30.

—— (1966), 'Tenure and Productivity of Philippine Rice Producing Farms', *Philippine Econ. J.*, 5(1), 42–63.

—— (1977), 'The Green Revolution: Seven Generalizations', *Inter. Devel. Rev.*, 19(1), 16–23.

Sanghavi, Prafulla (1969), *Surplus Manpower in Agriculture and Economic Development*. New York: Asia Publishing House.

Schickele, Rainer (1941), 'Effect of Tenure Systems on Agricultural Efficiency', *J. Farm Econ.*, 23(1), 185–207.

Schultz, Theodore W. (1940), 'Capital Rationing, Uncertainty, and Farm Tenancy Reform', *J. Polit. Econ.*, 48(3), 309–24.

—— (1964), *Transforming Traditional Agriculture*. New Haven, Conn.: Yale University Press.

Scott, James C. (1976), *The Moral Economy of the Peasant*. New Haven, Conn.: Yale University Press.

Scott, John T., Jr. (1970), 'Leasing Recommendations for Less-Developed Countries: An Extension of Leasing Theory', *Amer. J. Agr. Econ.*, 52(4), 610–13.

Sen, Abhijit (1981), 'Market Failure and Control of Labor Power: Towards an Explanation of "Structure" and Change in Indian Agriculture', Pts. 1 and 2, *Cambridge J. Econ.*, 5(3 and 4), 201–28 and 327–50.

References

Sen, Amertya K. (1966), 'Peasants and Dualism with or without Surplus Labor', *J. Polit. Econ.*, 74(5), 425–50.

—— (1975), *Employment, Technology and Development* Oxford: Clarendon Press.

Shaban, Radwan A. (1987), 'Testing between Competing Models of Sharecropping', *J. Polit. Econ.*, 95(5), 893–920.

Shahid, A., and Herdt, Robert W. (1982), 'Land Tenure and Rice Production in Four Villages of Dhaka District, Bangladesh', *Bangladesh Devel. Stud.*, 10(4), 113–24.

Shapiro, Carl, and Stiglitz, Joseph E. (1984). 'Equilibrium Unemployment as a Worker Discipline Device', *Amer. Econ. Rev.*, 74(3), 433–44.

Sharma, S. P. (1987), 'Agrarian Change and Agricultural Labor Relations: Nepalese Case Studies', Research Paper Series No. 4. HMG-USAID-GTZ-IDRC-Ford-Winrock Project. Kathmandu: Nepal.

Shavell, Steven (1979), 'Risk Sharing and Incentives in the Principal and Agent Relationship', *Bell J. Econ.*, 10(1), 55–73.

Shaw, Annapurna (1988), 'The Income Security Function of the Rural Sector: The Case of Calcutta, India', *Econ. Develop. Cult. Change*, 36(2), 303–14.

Shetty, Sudhir (1988), 'Limited Liability, Wealth Differences and Tenancy Contracts in Agrarian Economies', *J. Devel. Econ.*, 29(1), 1–22.

Shlomowitz, Rolph (1979), 'The Origins of Southern Sharecropping', *Agr. Hist.*, 53(3), 557–75.

Singh, Nirvikar (1989), 'Theories of Sharecropping', in Pranab K. Bardhan (1989), 33–72.

Smith, Adam (1937), *The Wealth of Nations*. New York: Modern Library Edition [1776].

Smith, Thomas C. (1959), *Agrarian Origins of Modern Japan*. Stanford, Calif.: Stanford University Press.

Spillman, William J. (1919), 'The Agricultural Ladder', *Amer. Econ. Rev.*, 9(1), Suppl. 170–9.

Stark, Oded, and Lucas, Robert E. B. (1988), 'Migration, Remittances, and the Family', *Econ. Develop. Cult. Change*, 36(3), 465–81.

Stewart, James C., and Arellano, Antonio B. (1975), 'Hagonoy, Davao del Sur', in *Changes in Rice Farming in Selected Areas of Asia*. Los Baños, Philippines: International Rice Research Institute, 303–24.

Stiglitz, Joseph E. (1974), 'Incentives and Risk Sharing in Sharecropping', *Rev. Econ. Stud.*, 41(2), 219–56.

—— (1975), 'Incentive, Risk Sharing, and Information: Notes toward a Theory of Hierarchy', *Bell J. Econ.*, 6(2), 552–79.

—— (1987), 'The Causes and Consequences of the Dependence of Quality on Price', *J. Econ. Lit.*, 25(1), 1–48.

—— (1988), 'Economic Organization, Information, and Development', in H. Chenery and T. N. Srinivasan, (eds.), *Handbook of Development Economics*. Amsterdam, Netherlands: North-Holland, 93–160.

Stiglitz, Joseph E. (1989), 'Rational Peasants, Efficient Institutions, and a Theory of Rural Organization: Methodological Remarks for Development Economics', in Bardhan (1989), 18–19.

—— and Weiss, Andrew (1981), 'Credit Rationing in Markets with Imperfect Information', *Amer. Econ. Rev.*, 71(3), 393–410.

Subbarao, K. (1987), 'Rapporteur's Report on Changing Structures of Ownership of Land and Associated Assets and Rural Labour Absorption in Different Regions', *Indian J. Agr. Econ.*, 42(3), 472–86.

Sutinen, J. G. (1975), 'The Rational Choice of Share Leasing and Implications for Efficiency', *Amer. J. Agr. Econ.*, 57(4), 613–21.

Takahashi, Akira (1969), *Land and Peasants in Central Luzon: Socio-Economic Structure of a Bulacan Village*. Tokyo, Japan: Institute of Developing Economies.

Talukder, R. K. (1980), 'Land Tenure and Efficiency in Boro Rice Production in an Area of Mymensingh District', *Bangladesh J. Agr. Econ.*, 3(2), 43–55.

Tamin, Moktar, and Mustapha, N. Hashim (1975), 'Kelantan, West Malaysia', in *Changes in Rice Farming in Selected Areas of Asia*. Los Baños, Philippines: International Rice Research Institute.

Taslim, M. A. (1988), 'Tenancy and Interlocking Markets: Issues and Some Evidence', *World Devel.*, 16(6), 655–66.

—— (1989), 'Short-Term Leasing, Resource Allocation, and Crop-Share Tenancy', *Amer. J. Agr. Econ.*, 71(3), 785–90.

Taylor, R. H. (1943), 'Post-Bellum Southern Rental Contracts', *Agr. Hist.*, 17(2), 121–8.

Thorner, Daniel, and Thorner, Alice (1962), *Land and Labor in India*. New York: Asia Publishing House.

Tongpain, S., and Jayasuriya, S. K. (1982), 'Tenancy, Farming Practices and Income Differences: A Study of Rice Farmers in Central Thailand'. Agricultural Economics Department Paper No. 82-12. Los Baños, Philippines: International Rice Research Institute.

Truran, James A., and Fox, Roger W. (1979), 'Resource Productivity of Land Owners and Sharecroppers in the Cariri Region of Ceara, Brazil', *Land Econ.*, 55(1), 93–107.

Umehara, Hiromitsu (1974), *A Hacienda Barrio in Central Luzon: A Case Study of a Philippine Village*. Tokyo: Institute of Developing Economies.

Utrecht, E. (1969), 'Land Reform in Indonesia', *Bulletin of Indonesian Econ. Stud.*, 5(3), 71–88.

Verma, B. N., and Bromley, D. W. (1987), 'The Political Economy of Farm Size in India: The Elusive Quest', *Econ. Develop. Cult. Change*, 35(4), 791–808.

References

Vyas, V. S. (1970), 'Tenancy in a Dynamic Setting', *Econ. Polit. Weekly*, 5(26), A73–80.

Walker, Thomas S., and Ryan, James G. (1990), *Village and Household Economies in India's Semi-Arid Tropics*. Baltimore: Johns Hopkins University Press.

Wallace, J., and Beneke, R. (1956), *Managing the Tenant-Operated Farm*. Ames, Ia.: Iowa State College Press.

Warr, Peter (1978), 'Share Contracts, Limited Information, and Production Uncertainty', *Australian Econ. Pap.*, 17(30), 110–23.

Warriner, Doreen (1969), *Land Reform in Principle and Practice*. London: Oxford University Press.

Weitzman, Martin L. (1986), 'Efficient Incentive Contracts', *Quart. J. Econ.*, 94(4), 719–30.

Wells, Mariam J. (1981), 'Social Conflict, Commodity Constraints, and Labor Market Structure in Agriculture', *Compar. Stud. in Society and Hist.*, 23(4), 679–704.

Williamson, Oliver E. (1975), *Market and Hierarchies. Analysis and Antitrust Implications*. New York: Free Press.

—— (1985), *The Economic Institutions of Capitalism*. New York: Free Press.

Wilson, Robert (1985), 'Reputations in Games and Markets', in Alvin E. Roth (ed.), *Game-theoretic Models of Bargaining*. Cambridge: Cambridge University Press, 27–62.

Winters, D. L. (1974), 'Tenant Farming in Iowa, 1860–1900: A Study of the Terms of Rental Leases', *Agr. Hist.*, 48(1), 130–50.

Yaari, Menaham E. (1965), 'Uncertain Lifetime, Life Insurance, and the Theory of the Consumer', *Rev. Econ. Stud.*, 2(2), 137–50.

Zaman, M. Raquibuz (1973), 'Sharecropping and Economic Efficiency in Bangladesh', *Bangladesh Econ. Rev.*, 1(2), 149–72.

Index of Names

Abdullah, A. 103
Abraham, A. 32 n.
Adams, D. W. 4, 29, 79
Akash, M. M. 103
Akerlof, G. A. 86
Alchian, A. A. 1, 5
Allen, F. 43–4, 49, 114
Alston, L. J. 27, 29, 35, 45,
  55 n., 98, 100, 101
Anderson, J. N. 150
Anker, R. 12
Aoki, M. 178
Appu, P. S. 87
Arellano, A. B. 99
Arrow, K. J. 3, 5, 32 n., 72, 124,
  144

Bagi, F. S. 92
Banzon, L. E. 105
Bardhan, K. 24
Bardhan, P. K. 2, 4, 5, 16, 17,
  21, 22, 23, 24, 25, 27, 29, 31,
  35, 36, 55 n., 63, 70, 74, 77,
  79, 81, 86, 97, 151, 171 n., 173
Barker, R. 15, 96, 150, 155
Basant, R. 86, 151, 171 n.
Basu, K. 74, 105
Bauzon, L. E. 87
Becker, G. S. 17
Bell, C. 2, 4, 5, 16, 22, 24, 25,
  27, 29, 30, 31, 33 n., 35,
  40–1, 44, 54 n., 55 n., 63, 70,
  75–6, 78, 81, 84 n., 94, 102,
  103, 106 n., 109, 176
Ben-Porath, Y. 100, 125
Beneke, R. 45

Benoit, J.-P. 68 n.
Berry, R. A. 14
Berry, R. L. 77, 79, 81
Bester, H. 73
Bhaduri, A. 74, 95, 147 n.
Bhagwati, J. N. 4, 14, 29
Bhalla, S. S. 14, 24, 86
Bharadwaj, K. 44, 92, 100
Bharathan, K. 87
Bhuiyan, M. S. R. 102
Binswanger, H. P. 2, 13, 16, 23,
  33 n., 59, 70, 83 n., 86,
  171 n.
Bliss, C. J. 3, 15, 30, 37, 79, 80,
  81, 83 n., 84 n., 99, 101, 102,
  105 n.
Bradly, M. E. 5
Braverman, A. 5, 22, 27, 29, 30,
  31, 32 n., 33 n., 54 n., 71, 72,
  73, 74, 75, 76, 80, 83 n.,
  84 n., 96
Breman, J. 24, 70, 86, 101
Brewster, J. M. 13, 100
Bromley, D. W. 9, 14
Bull, C. 26, 58

Cain, M. 102
Caldwell, J. C. 12
Carmichael, H. L. 32 n., 58
Carter, M. R. 55 n., 83 n.
Castillo, G. T. 70, 97
Castle, E. 4, 29, 78
Chakravarty, A. 106 n.
Chakravarty, J. 25
Chakravarty, S. 14
Chandra, N. K. 92

Index of Names

Chao, K. 77, 97
Chayanov, A. V. 7
Cheung, S. N. S. 3, 4, 21, 23, 27, 30, 36–9, 49, 77, 81, 83 n., 97, 173
Chuma, H. 62
Clark, M. G. 5
Clayton, E. S. 9
Cline, W. R. 14
Coase, R. H. 1, 5, 36
Cohen, P. T. 44, 100
Cordova, V. G. 150
Currie, J. M. 27, 35, 79

Dantwala, M. L. 87, 101
Das, P. K. 44, 100
Datta, S. K. 31, 45, 74, 173
David, C. C. 96
Day, R. J. 101
Demsetz, H. 1, 5
Dow, N. 44
Dowell, R. S. 97, 98
Drake, L. S. 4, 29

Ely, R. T. 81, 99
Embree, J. F. 99
Eswaran, M. 2, 29, 54 n., 59–60, 62, 64, 66–7, 68–9 n., 77, 172, 173, 176, 177

Feder, G. 68 n., 83 n., 96, 177
Ferrie, J. P. 100
Finkler, K. 70, 99
Fox, R. W. 81, 92
Friedman, M. 51
Fujimoto, A. 25, 77, 81, 87, 97, 100, 123
Furnival, J. A. 109

Galpin, C. J. 81, 99
Gangopadhyay, S. 29, 33 n., 74, 84 n.
Gapud, J. P. 70
Geertz, C. 16, 109

George, A. 86
Georgescu-Roegen, N. 4, 29
Ghose, A. K. 151, 171 n.
Green, J. R. 32 n.
Guasch, J. L. 54 n., 83 n.
Guhan, S. 87

Hallagan, W. 41–3
Harris, M. 21, 35
Hart, G. 70
Hart, O. 3, 16, 32 n., 33 n.
Hayami, Y. 2, 5, 9, 14, 17, 22, 24, 25, 30, 44, 70, 86, 87, 97, 99, 100, 101, 105, 108, 109, 123, 127 n., 133, 149, 150, 154, 156, 158
Heady, E. O. 4, 29, 78, 92, 97
Hendry, J. B. 77
Herdt, R. W. 15, 88, 96, 102, 149, 155
Herring, R. J. 19, 87
Hester, E. D. 150
Hiebert, L. D. 29, 49
Higgs, H. 25
Higgs, R. 55 n., 97, 98, 99
Hirashima, S. 15, 68 n., 86, 87, 105
Hirschmeir, J. 178
Hirshleifer, J. 48
Ho, S. P. S. 32 n., 77
Hoffman, P. T. 97, 99
Holmstrom, B. 3, 21, 26, 32 n., 33 n., 35, 58, 78
Horii, K. 77
Hossain, M. 102, 103
Hsiao, J. C. 36
Huang, Y. 97
Hurwicz, L. 29
Husken, F. 70, 81, 86

Issawi, C. 4, 29, 78

Jabbar, M. A. 25, 102
James, W. T. 25, 70

Index of Names

Jannuzi, F. T. 87
de Janvry, A. 9, 100
Jayasuriya, S. K. 97
Jaynes, G. D. 3, 27, 33 n., 79, 80, 83 n.
Jewitt, I. 33 n.
Jodha, N. S. 70, 83 n., 102
Johnson, D. G. 30, 33 n., 38, 97
Joshi, P. C. 106 n.
Junankar, P. N. 92

Kawagoe, T. 14
Khandker, S. R. 12
Khasnabis, R. 25
Khusro, A. M. 87
Kikuchi, M. 17, 24, 70, 86, 97, 99, 100, 101, 105, 108, 109, 123, 149, 150, 156, 158
Kislev, Y. 7
Kmenta, J. 147 n.
Knowles, K. C. 12
Koo, A. Y. C. 19, 27, 29
Kotwal, A. 2, 28, 29, 54 n., 59–60, 62, 64, 66–7, 68–9 n., 77, 86, 172, 173, 176, 177
Krishna, V. 68 n.
Kuchiba, M. 87, 105
Kurz, N. 72
Kutcher, G. P. 92

Ladejinsky, W. 9, 19, 87, 101, 102
Lambert, R. A. 57
Lau, L. J. 14
Ledesma, A. J. 96, 150
Lehmann, D. 99
Leibenstein, H. 178
Levinthal, D. 3, 33 n.
Lewis, W. A. 4, 5
Lin, J. Y. 5
Lowe, R. 9
Lucas, R. E. B. 12, 29, 35, 68 n.
Luce, R. D. 68 n.

Mabun, P. 150

McIntire, J. 16, 33 n., 70
McLennan, M. S. 150, 158
McMillan, R. T. 150
Mandal, M. A. S. 102
Mangahas, M. 19, 25, 27, 29, 70, 77, 96, 171 n.
Marcos, F. 149
Marshall, A. 3–4, 30, 32 n., 37, 38, 70
Marx, K. 133
Mazumdar, D. 4, 29, 35
Milgrom, P. 32 n.
Mill, J. S. 3
Mirrlees, J. A. 4, 56
Mitra, P. K. 22, 29, 32 n., 33 n., 47, 71, 73, 74, 75
Mitrany, D. 133
Moffet, D. 72
Morooka, Y. 25, 44, 109, 127 n.
Mundlak, Y. 15
Murakami, N. 32 n., 48, 55 n., 147 n.
Murrell, P. 45, 67, 69 n., 78, 138, 173
Myerson, R. B. 32 n., 78

Nabi, I. 70, 79, 97, 98, 124
Nandan, D. S. 102
Narain, D. 106 n.
Nelson, J. M. 12
Newbery, D. M. G. 2, 23, 28, 29, 33 n., 48, 49, 54 n., 79, 96
Nugent, J. B. 31

Oishi, S. 89
Otsuka, K. 2, 15, 22, 30, 32 n., 48, 52, 55 n., 62, 68 n., 88, 96, 145 n., 147 n., 149, 155, 159, 171 n.

Pal, T. K. 25, 96
Pant, C. 27, 29
Parthasarathy, G. 70, 81, 96
Peach, J. T. 87

Peterson, W. 7
Platteau, J.-P. 32 n.
Pollak, R. 100, 125
Posner, R. A. 12
Prasad, D. S. 70, 81, 96
Prosterman, R. L. 19, 106 n.

Quezon, M. 149
Quibria, M. G. 2, 27, 29

Radner, R. 4, 57
Raiffa, H. 68 n.
Raj, K. N. 92, 100
Rao, C. H. H. 25, 29, 79, 94, 97, 99, 100, 109, 119
Rao, J. M. 27, 29
Rashid, S. 2, 27, 29
Rask, N. 4, 29, 79
Raviv, A. 21, 35
Ray, D. 82
Reid, J. D. 33 n., 48, 81, 97, 104, 116, 143
de los Reyes, B. N. 154
Richards, A. 24
Riedinger, J. M. 19, 106 n.
Riley, J. G. 48
Rivera, G. E. 150
Robertson, A. F. 9, 99
Rosenzweig, M. R. 2, 12, 13, 16, 23, 33 n., 59, 70, 83 n., 97, 106 n.
Roumasset, J. A. 2, 15, 24, 25, 36, 70, 79, 80, 92, 100, 105
Roy, P. 14
Rubinstein, A. 4, 57
Rudra, A. 24, 25, 70, 81, 86, 106 n., 151
Russel, R. 147 n.
Ruttan, V. W. 14, 19, 96
Ryan, J. G. 102

Sanghavi, P. 24, 86
Scandizzo, P. L. 92

Schickele, R. 4, 29, 70, 78
Schultz, T. W. 70, 98, 105
Scott, J. C. 12, 17, 70
Scott, J. T. 29
Sen, A. 79
Sen, A. K. 4, 14, 29
Sengupta, K. 29, 33 n., 74, 82, 84 n.
Shaban, R. A. 102, 103
Shah, C. H. 87, 101
Shahid, A. 102
Shapiro, L. 29, 68 n.
Sharma, S. P. 87
Shavell, S. 21, 35
Shaw, A. 12
Shetty, S. 29, 77
Shlomowitz, R. 25, 70
Siller, D. A. 83 n.
Singh, N. 2, 68 n., 79
Smith, A. 3
Smith, T. C. 89
Spillman, W. J. 2, 64, 104, 151
Srinivasan, T. N. 2, 4, 5, 21, 23, 25, 27, 29, 32 n., 33 n., 36, 63, 70, 71, 72, 75–6, 81
Stark, O. 12
Stern, N. H. 3, 15, 30, 37, 79, 80, 81, 83 n., 84 n., 99, 101, 102, 105 n.
Stewart, J. C. 99
Stiglitz, J. E. 2, 4, 21, 22, 23, 24, 25, 27, 29, 30, 33 n., 47, 48, 49, 54 n., 68 n., 71, 73, 74, 75, 80, 84 n., 96
Stokey, N. L. 32 n.
Subbarao, K. 101
Sussagkarn, C. 44
Sutinen, J. G. 48

Takahashi, A. 99, 150
Talukdar, R. K. 102
Taslim, M. A. 54 n., 83 n.
Taylor, R. H. 81

Index of Names

Thorner, A. 24, 70, 86, 101
Thorner, D. 24, 70, 86, 101
Tongpain, S. 97
Truran, J. A. 81, 92

Umehara, H. 97, 99, 150
Utrecht, E. 108
Uy, M. 2, 15, 24, 100, 105

Verma, B. N. 14
Vyas, V. S. 25, 81

Walker, T. S. 102
Wallace, J. 45
Warr, P. 35

Warriner, D. 19
Weiss, A. 73
Wells, M. J. 25
Williamson, O. E. 1, 5
Wilson, R. 57, 69 n.
Winters, D. L. 99
Wolpin, K. J. 106 n.

Yaari, M. E. 4, 57, 68 n., 72
Yotopoulos, P. A. 14
Yui, T. 178

Zaman, M. R. 25
Zusman, P. 4, 27, 29, 35, 40–1, 54 n., 63, 176

Index of Subjects

Africa 6–9, 12
agency:
   problem described 21–2
   theory 3, 35
agrarian economies:
   data 6, 85–105, 149–50
   described 6–18
   stratified population 1
   structures 6–12
'agricultural ladder' 104–5, 116, 143, 151, 158, 159, 176
   see also landlords; tenants
altruistic behaviour 131–3
altruistic-community hypothesis 133
Asia 7–9, 12

Bangladesh 86, 87, 101–3, 170

cash crops 108–9, 119
casual labour 15, 60, 163–4
Chile 24
contract choice 12, 172–3
   and efficiency 103–4
   optimum 34–54
   restricted 174–5
contract enforcement 46–8, 53–4, 133, 135, 173
   conditions of 141–3
   controversy on 4–5
   costs 12–13, 117, 120–3, 166
   landlord supervision 123–4
   mechanism of 98–101
   unenforceable 34, 49–52, 74–6, 135
   work-effort 28–31, 49, 74–6, 85

   see also personal ties; reputation
contracts:
   basic model 21–31
   cost-sharing 78–81
   credit 71–8, 114
   as insurance 77
   interlinked 70–83
   long-term 56–69, 77
   pawn 114, 157, 159
   rental 128, 130–1
   subtenancy 156
   verification of breach of 56
   see also fixed-rental contracts; permanent labour contracts; share contracts; and under Jeepnies
credit arrangements 71–8, 114, 123
credit market 177
   imperfect 70, 82
crop insurance 16

economies of scope 76
   see also under scale
efficiency:
   and contract choice 103–5
   hypothesis 36–9
   of owner and tenants 31
   relative 117–23
   tests 137–41
   under share tenancy 89–98
   see also inefficiency
efficient-community hypothesis 134, 135, 141, 144
Egypt 24

Index of Subjects

elasticity of substitution,
  estimation of 145
employers, characteristics
  of 161–2
enforcement, see contract
  enforcement; fixed-rent
  contracts; share contracts
equivalence theorem 49
Europe 7

farm size 5, 7–12, 14, 99, 176–7
farm-helper 150–1, 153
farmers, characteristics of 160–1
fertilizer input 94–5
fixed-rent contracts 3–4
  enforcement costs 99
  prevalence of 66–7
  shift from share contracts 128, 134–5
fixed-wage contract 66
France 25

income, tenant 97–8
income distribution:
  by contract type 141, 144
  under share tenancy 38–9, 48, 50, 51, 89–98, 101
income redistribution 158–9
India 12, 14, 24, 74, 75, 86–8, 101–3, 104, 159, 170
Indonesia 19, 81, 86, 87, 107–27
  agrarian law 108
  cash crops 108–9
  land-tenure systems 113–17
  study village:
    characteristics 109–13;
    farming systems 112–13;
    landholdings 110–12
inefficiency 34–5, 37–9, 101
  see also efficiency; Marshallian
    theory
inefficient-community
  hypothesis 134, 135

information 58
innovation speed 95–7
input complementarity 71
insurance:
  crop 16
  long-term contract as 77
  interest rate 121
  see also under contracts,
    interlinked

Japan 14
  agrarian history 89
  management 178
Jeepnies 128–47
  contract distribution 131, 133
  contract enforcement 128, 130–1, 141
  efficiency tests 137–41
  income distribution 141, 144
  mode of operations 129–30
  owner-drivers 138
  owners association 129–30
  ownership 129
  rental contracts 128, 130–1
  survey 135–7

labour:
  casual 2, 15, 22, 60, 163–4
  family versus permanent 164–6
  landless 160–1
  and legislative restraints 175
  semi-attached 151, 156
  short-term 150
  see also permanent labour
land legislation 86, 170, 175
land reform 87–8, 108, 154, 158
land-pawning 157, 159
landlord–tenant relationship 9
landlords:
  advantage as creditor 76–7
  characteristics 115–16
  optimization problem 35, 36, 38, 46

landlords (*cont.*):
  utility 26, 62
Latin America 7–9, 12

Malaysia 86, 87
management ability 65
market, and community 16–18
Marshallian theory 29, 34–5, 37–9, 41, 78–80, 131–2, 173–4
model:
  bargaining 40–1
  basic 18, 21–31; formulation of the optimization problem 26–31; specification of functional forms 22–6
  land-quality transaction-cost 45
  self-selection 41–4

Nepal 86, 87
North America 7, 25

optimization problem:
  landlord's 35, 36, 38, 46
  worker's 61
optimum conditions, derivation 52–3
optimum contract choice 34–54
output, by contract type 117–19
output-sharing 24–5
owners:
  cultivation 7, 113
  efficiency 31
  Jeepney 129–30, 138
  urban residents 156–8

Pakistan 86–8
patron–client relationship 17
peasants 7
penalty 56
permanent labour 2, 7, 15–16, 24, 148
  characteristics of 161
  conditions of 85–9
  income 163–4
  long-term contracts 59–64
  new 151–4
  and profit levels 166–70
  spread of 154, 155
  traditional 150–1
  used by big landlords 156–8
  versus family labour 164–6
  work-effort of 21–2
personal ties 17–18, 125
  Jeepney contracts 128, 133–4, 141
  scale diseconomies 143
  see also reputation
Philippines 12, 20, 86, 87, 88, 99, 101
  change in labour relations 150–4
  Jeepnies 128–47
  Rice Bowl 148–71
  study area described 148–9
  surveys 149–50
production, and labour type 166–70
production function, *see* scale
profit, and labour type 166–70

rent:
  comparisons 97–8, 120–3
  levels 175
  paid in advance 114
reputation 57–8, 59, 64–7, 78, 126, 173
  see also personal ties
resource allocation 38, 48, 51
reward function 22
rice technology 155
risk 49, 133–4
risk premium 48, 123, 144
risk-aversion 29, 34, 39, 46, 51, 172–3
  Jeepnies 141
  and share-tenancy 64–5

Index of Subjects

risk-neutral 172–3
risk-sharing 12, 22, 48–51, 63, 85, 144
scale:
  constant returns 14, 23
  economies of 13–15, 49
self-enforcing mechanism 56, 58
self-selection 22, 41–4
share contracts 3–4, 16–17, 19, 24
  abolition of 88
  effective enforcement 98–101
  efficiency 89–98, 101
  50:50 sharing 41, 78, 80, 81, 82, 174
  income distribution 89–98
  and risk-aversion 64–5
  shift to fixed-rental 128, 134–5
  under certainty 39–45
share-cropping 2, 114
social interactions, theory of 17
soil fertility depletion 45, 67
squatters 7

technology 95–7, 155

tenants:
  characteristics 115–16
  efficiency 31
  work-effort of 21–2
tenure:
  and farm size 7–12
  tribal or communal 7
Thailand 86, 87
theoretical conclusions 172–4
transaction costs 17, 49, 144, 156
  see also contract enforcement

uncertainty 29, 34
United States 98
utility function 25–6

work-effort 23
  enforceability 28–31, 49, 74–6, 85
  of tenants and permanent labourers 21–2
work-incentives 5, 22, 173–4
worker's optimization problem 61

yields 90–7

DATE DUE

FEB –	1994		
	261-2500		Printed in USA